EMA 2024: Basic Science and Electronics Division Meeting

Denver, Colorado, USA
13-16 January 2024

ISBN: 978-1-7138-9235-9

Printed from e-media with permission by:

Curran Associates, Inc.
57 Morehouse Lane
Red Hook, NY 12571

Some format issues inherent in the e-media version may also appear in this print version.

Printed by Curran Associates, Inc. (2024)

For permission requests, please contact the American Ceramic Society
at the address below.

The American Ceramic Society
550 Polaris Parkway, Suite 510
Westerville, Ohio 43082

Phone: 1-866-721-3322 (toll free U.S. only)
1-614-890-4700 (outside U.S.)
Fax: 1-614-899-6109

ceramics.org

Additional copies of this publication are available from:

Curran Associates, Inc.
57 Morehouse Lane
Red Hook, NY 12571 USA
Phone: 845-758-0400
Fax: 845-758-2634
Email: curran@proceedings.com
Web: www.proceedings.com

TABLE OF CONTENTS

Found 295 Records

CONTROL ID: 3959852
CURRENT SYMPOSIUM: S7 - Defects and Transport in Ceramics
CURRENT SESSION: Predictive bulk and interfacial point defect energetics and equilibria from density functional theory, molecular dynamics, and other computational methods
TITLE: In silico screening of effective dopants for improving insulation resistance in $BaTiO_3$
PRESENTATION TYPE: Contributed (Oral)
PRESENTER (FIRST NAME ONLY): Tomosato
PRESENTER (LAST NAME ONLY): Kanagawa
PRESENTER (INSTITUTION ONLY): Murata Manufacturing Co., Ltd.
PRESENTER (EMAIL ONLY): tomosato.kanagawa@murata.com
PRESENTER (COUNTRY ONLY): Japan
DATE/TIME SUBMITTED: September 21, 2023, 10:18 PM
ABSTRACT BODY:
Abstract Body: Understanding the effects and solubility of dopants is crucial for improving material properties, but it is challenging to clarify them experimentally. Computational simulations are considered effective for this purpose. In this study, we developed a calculation model for equilibrium defect concentrations in materials containing various dopants by considering the solubility. Conventional calculations use dopant concentrations as parameters, which do not consider dopant solubility and make it difficult to properly evaluate defect and carrier concentrations. Our model considers the concentrations of competitive phases formed by segregated dopants and uses the sum of the concentrations of the dopant dissolved into the lattice and competitive phases as a parameter. In addition, by fixing the defect concentrations with a high migration barrier with respect to temperature, we considered the thermal history of the processes used in the experiments. We applied this model to $BaTiO_3$ and conducted systematic calculations of the defect concentrations for over 50 dopants in $BaTiO_3$. The results revealed that V and Mo can simultaneously reduce electron/hole carrier and oxygen vacancy concentrations, suggesting that they are effective dopants for enhancing insulation resistance in $BaTiO_3$ sintered in a reducing atmosphere. These results would significantly contribute to more reliable macroscale simulations.

CONTROL ID: 3964660
CURRENT SYMPOSIUM: S18 - Structure and function in high-entropy, entropy-stabilized, and related oxides
CURRENT SESSION: Characterization and quantification of structure and properties
TITLE: Spin, charge, and lattice degrees of freedom in high entropy oxides
PRESENTATION TYPE: Contributed (Oral)
PRESENTER (FIRST NAME ONLY): Alessandro
PRESENTER (LAST NAME ONLY): Mazza
PRESENTER (INSTITUTION ONLY): Los Alamos National Lab
PRESENTER (EMAIL ONLY): armazza@lanl.gov
PRESENTER (COUNTRY ONLY): United States
DATE/TIME SUBMITTED: September 25, 2023, 11:39 AM
ABSTRACT BODY:
Abstract Body: The multitude of properties observed in strongly correlated materials are deeply intertwined with the spin, charge and lattice degrees of freedom. In high entropy oxides, it has been shown that these degrees of freedom are highly tunable – from a uniform to a highly frustrated state. Using the case study of the ABO_3 perovskite $La_xSr_{1-x}(Cr_{0.2}Mn_{0.2}Fe_{0.2}Co_{0.2}Ni_{0.2})O_3$ we demonstrate how, by having a large variance in the degrees of freedom site to site, high entropy materials can extend beyond known phase diagrams and exhibit unique properties beyond their parent compounds. For the first time, we explore in compliment to the local spin and lattice changes with doping, the effect on charge and charge compensation in this system. This provides first of its kind insight into the electronic character of high entropy materials. These results supply a blueprint for understanding valence in high entropy oxides and demonstrate how variance is a critical order parameter in the exploration of correlated oxides, yielding continuous tunability to charge, electronic, structural, and magnetic orderings extending beyond known phase diagrams.

CONTROL ID: 3965282

CURRENT SYMPOSIUM: S11 - In situ/operando Characterization of Nanomaterials

CURRENT SESSION: Study of electronic materials such as Mott insulators, dielectrics, ionics, ferroics, semiconductors, superconductors etc. using advanced characterization techniques

TITLE: Mapping the Domain Structure in Bicrystal Lead Zirconate-Titanate thin films using Nanodiffraction

PRESENTATION TYPE: Contributed (Oral)

PRESENTER (FIRST NAME ONLY): Stanislav

PRESENTER (LAST NAME ONLY): Udovenko

PRESENTER (INSTITUTION ONLY): Pennsylvania State University

PRESENTER (EMAIL ONLY): spu5065@psu.edu

PRESENTER (COUNTRY ONLY): United States

DATE/TIME SUBMITTED: September 16, 2023, 11:08 AM

ABSTRACT BODY:

Abstract Body:

Domain wall motion in ferroelectric materials is a significant contributor to the functional properties, including the dielectric and piezoelectric responses. However, domain wall motion in perovskite ferroelectrics is often restricted by grain boundaries. Thus, quantification of the degree of domain wall pinning near grain boundaries is important for optimizing the properties of ferroelectric materials. Here, the first experimental results collected using nanodiffraction at the Hard X-Ray Nanoprobe beamline at NSLS-2 (Brookhaven National Lab) are reported. For this work, 200 nm $PbZr_{0.2}Ti_{0.8}O_3$ (PZT20/80) films were deposited on $SrRuO_3$/$SrTiO_3$ bicrystal substrates, with 5 μm wide stripe electrodes to apply DC electric field to the grain boundary area. Nanobeam diffraction was then utilized to record a set of diffraction patterns from 6×5.1um area at increasing applied field.

The experimental data shows that on increasing the electric field amplitude, the mean area of domains with polarizations along electric field expands, while the areas of domains with polarizations perpendicular to electric field shrinks. This was confirmed both by tracking evolution of single domain and by statistical analysis of the mean domain sizes. Using statistical analysis of domains geometrical parameters it was found that changes in the domain shapes are reduced in the vicinity of the grain boundary.

CONTROL ID: 3965690
CURRENT SYMPOSIUM: S11 - In situ/operando Characterization of Nanomaterials
CURRENT SESSION: Real-time investigation of mechanical properties in structural materials.
TITLE: Measuring Microstructural and Micromechanical Response during in situ Loading with Diffraction Microstructure Imaging
PRESENTATION TYPE: Invited (by Invitation Only)
PRESENTER (FIRST NAME ONLY): Darren
PRESENTER (LAST NAME ONLY): Pagan
PRESENTER (INSTITUTION ONLY): Pennsylvania State University
PRESENTER (EMAIL ONLY): dcp5303@psu.edu
PRESENTER (COUNTRY ONLY): United States
DATE/TIME SUBMITTED: July 13, 2023, 09:18 AM
ABSTRACT BODY:
Abstract Body: This presentation will provide an introduction and overview of diffraction microstructure imaging (DMI) techniques, with a focus on high-energy X-ray diffraction microscopy (HEDM). HEDM is a class of monochromatic, angular dispersive diffraction techniques which provide a unique means to non-destructively probe 3D microstructural and micromechanical response of polycrystalline materials at grain and sub-grain scales, enabling in situ and in operando studies. While these techniques have been primarily developed for studying engineering alloys, they naturally extend to other polycrystalline systems, including electronic and geological materials. A range of HEDM in situ case studies demonstrating different thermomechanical applications, combinations with material modeling, and extension beyond engineering alloys will be discussed.

CONTROL ID: 3968315

CURRENT SYMPOSIUM: S11 - In situ/operando Characterization of Nanomaterials

CURRENT SESSION: In operando study of device performances and behaviors including batteries, fuel cells, solar cells, memristors, memory devices, etc.

TITLE: Nanoscale In-Operando Characterization of Buried Interface Structures in Batteries and Related Material Using Neutron Reflectometry

PRESENTATION TYPE: Invited (by Invitation Only)

PRESENTER (FIRST NAME ONLY): Joseph

PRESENTER (LAST NAME ONLY): Dura

PRESENTER (INSTITUTION ONLY): National Institute of Standards and Technology

PRESENTER (EMAIL ONLY): Joseph.Dura@nist.gov

PRESENTER (COUNTRY ONLY): United States

DATE/TIME SUBMITTED: August 23, 2023, 03:20 PM

ABSTRACT BODY:

Abstract Body: Neutron reflectometry, NR, is ideally suited to investigate battery and fuel cell interfaces in both liquid and solid electrolyte systems. This is because NR provides a depth profile of the Scattering Length Density, a parameter of materials derived from composition, with Angstrom level precision and accuracy for features on the order of a few nanometers up to ~500nm. Because of the penetrating power of neutrons, NR can probe complex sample environments, including model batteries, providing in operando characterization of unperturbed interfaces in their native environment under applied potential. Unlike other probes such as x-rays and electrons, where the scattering potential is proportional to the atomic number, neutrons interact with the nucleus, so the scattering potential can be relatively large for light elements and can differ greatly for isotopes of some elements such as Li and H. I will introduce NR, discuss its unique capabilities, and give examples of how it can be applied to investigate battery and fuel cell interfaces, e.g., to characterize battery electrode and SEI layers as a function of potential and cycling, and to investigate interfacial structures in Nafion and other fuel cell materials.

CONTROL ID: 3970606
CURRENT SYMPOSIUM: S10 - Fundamental mechanisms in materials for next generation computing devices
CURRENT SESSION: Fundamental materials needs for emerging computing paradigms
TITLE: Engineering of Dynamic Behaviors in Emerging Memory Devices for Neuromorphic Circuits
PRESENTATION TYPE: Invited (by Invitation Only)
PRESENTER (FIRST NAME ONLY): Karsten
PRESENTER (LAST NAME ONLY): Beckmann
PRESENTER (INSTITUTION ONLY): NY CREATES
PRESENTER (EMAIL ONLY): kbeckmann@sunypoly.edu
PRESENTER (COUNTRY ONLY): United States
DATE/TIME SUBMITTED: September 25, 2023, 11:06 AM
ABSTRACT BODY:
Abstract Body: The last two decades showed an increased interest in emerging two -terminal memory devices and has sparked a significant global research effort. Technological implementations include ionic devices utilizing anion or cation movement in dielectrics, ferroelectric films embedded in various device structures and phase change materials among others. All these technologies have been shown to exhibit analog changes of their respective state and were subsequently proposed to be implemented into neuromorphic architectures as synapses or neurons. It is essential to understand the dynamic behavior of these novel devices to either evaluate their performance or utilize them in neuromorphic architectures. First, this presentation will review the time ranges of the dynamic behavior based on the underlying physical properties and connect those to the requirements of demonstrated or postulated neuromorphic architectures and circuits. Second, it will outline engineering approaches to tune the dynamic behavior of selected technologies to either be used as synapses or neurons.

CONTROL ID: 3971061

CURRENT SYMPOSIUM: S11 - In situ/operando Characterization of Nanomaterials

CURRENT SESSION: Study of electronic materials such as Mott insulators, dielectrics, ionics, ferroics, semiconductors, superconductors etc. using advanced characterization techniques

TITLE: Characterizing phase transitions in strongly-correlated oxides at atomic resolution with continuously variable temperature cryo-STEM

PRESENTATION TYPE: Invited (by Invitation Only)

PRESENTER (FIRST NAME ONLY): Noah

PRESENTER (LAST NAME ONLY): Schnitzer

PRESENTER (INSTITUTION ONLY): Cornell University

PRESENTER (EMAIL ONLY): nis29@cornell.edu

PRESENTER (COUNTRY ONLY): United States

DATE/TIME SUBMITTED: August 18, 2023, 04:18 PM

ABSTRACT BODY:

Abstract Body: Cryogenic scanning transmission electron microscopy (cryo-STEM) provides atomic-resolution structural insight at the low temperatures required to access strongly-correlated behavior exhibited by many functional ceramics. High-resolution cryogenic measurements, however, have conventionally been limited to the temperature set by the cryogen boiling point, limiting access to phases stable over narrow temperature ranges and dynamic temperature-dependent processes such as phase transitions. Here, we leverage a novel liquid-nitrogen-cooled STEM holder design integrating MEMS devices which locally heat the sample without destabilizing the system to build a microscopic understanding of a low temperature metal-insulator transition (MIT).

Ca_2RuO_4 is a Mott insulator with an MIT at 357 K in the bulk. The MIT is extremely sensitive to stimuli including electric fields and strain, and in epitaxial thin films grown on $LaAlO_3$ the MIT is suppressed to ~230 K. Cooling through the MIT we measure at atomic resolution with cryo-STEM an emergent strain-mediated nanostructure of coexisting phases and characterize its evolution as the film is cycled through the MIT. These measurements reveal new complexity in the insulating phase of these strained films, and demonstrate the additional insight provided by continuous cryogenic temperature control in the STEM.

CONTROL ID: 3971706
CURRENT SYMPOSIUM: S9 - Machine Learning and Automated Synthesis/Characterization for Novel Materials
CURRENT SESSION: Autonomous research
TITLE: Accelerated Materials Discovery Enabled by an End-to-End Autonomous Fluidic Lab
PRESENTATION TYPE: Invited (by Invitation Only)
PRESENTER (FIRST NAME ONLY): Milad
PRESENTER (LAST NAME ONLY): Abolhasani
PRESENTER (INSTITUTION ONLY): North Carolina State University
PRESENTER (EMAIL ONLY): abolhasani@ncsu.edu
PRESENTER (COUNTRY ONLY): United States
DATE/TIME SUBMITTED: August 07, 2023, 12:27 AM
ABSTRACT BODY:
Abstract Body: Despite the intriguing size-dependent properties and widespread applications of semiconductor nanocrystals in energy and chemical technologies, their discovery and development are still time-, resource-, and labor-intensive. Recent advances in flow chemistry and autonomous experimentation provide an exciting opportunity to accelerate the discovery and development of semiconductor nanocrystals from 10+ years to less than 1 month. In this talk, I will present an end-to-end self-driving fluidic lab for accelerated discovery and development of semiconductor nanocrystals, through integration of flow chemistry, colloidal nanoscience, and in-situ characterization with data science. I will discuss how standardization and modularization of the hardware of the self-driving fluidic labs can enable data intensification (>100 experiments/day) while minimizing chemical consumption to achieve fast-tracked navigation through high dimensional experimental design space of energy materials (>10^{20} possible experimental conditions). Example applications of the end-to-end self-driving fluidic lab for the synthetic route discovery of energy materials will be presented. Finally, I will present a fully autonomous fluidic lab operated by reinforcement learning to rapidly discover the multi-step chemistry of hetero-nanostructures with no prior knowledge within 30 days of continuous operation.

CONTROL ID: 3972952
CURRENT SYMPOSIUM: S11 - In situ/operando Characterization of Nanomaterials
CURRENT SESSION: In operando study of advanced functional materials including transition metal oxides, graphene, transition metal dichalcogenides, halide perovskites, etc.
TITLE: Active learning for physics discovery in autonomous microscopy
PRESENTATION TYPE: Contributed (Oral)

PRESENTER (FIRST NAME ONLY): Yongtao
PRESENTER (LAST NAME ONLY): Liu
PRESENTER (INSTITUTION ONLY): Oak Ridge National Lab
PRESENTER (EMAIL ONLY): liuy3@ornl.gov
PRESENTER (COUNTRY ONLY): United States

DATE/TIME SUBMITTED: September 11, 2023, 12:10 PM

ABSTRACT BODY:

Abstract Body: In this work, we developed deep kernel learning (DKL)-driven and deep convolutional neuron network (DCNN)-driven autonomous scanning probe microscopy (SPM) to investigate the structure-property relationship in ferroelectric materials and photovoltaic perovskites. First, we implemented DKL, an active learning model, that actively learns the relationship between structural elements in image data and properties encoded in spectroscopic data during real-time microscopy experiments. We showed the application of this method in ferroelectric materials and hybrid perovskites. DKL-SPM revealed larger polarization mobility near c-/c+ domain walls in a $PbTiO_3$ thin film and anomalous conductivity around the ground boundary junction points of a hybrid perovskite thin film. Second, we implemented an ensembled DCNN in SPM, which converts on-the-fly SPM images into semantically segmented images for objects of interest, then SPM will perform spectroscopic measurements on these objects to investigate their detailed properties in an automated manner. This approach unveiled that the grain boundary junction points exhibit the lowest conductivity. It also reveals a distinct polarization magnitude difference of adjacent ferroelastic domain walls in the $PbTiO_3$ film. We anticipate the established approaches here can be extended to microscopy methods other than SPM in the future to accelerate materials and physics discovery.

CONTROL ID: 3975129
CURRENT SYMPOSIUM: S10 - Fundamental mechanisms in materials for next generation computing devices
CURRENT SESSION: Electrochemical and reservoir memory and computing devices
TITLE: Design of Memristive Devices Towards Neuromorphic Computing
PRESENTATION TYPE: Contributed (Oral)
PRESENTER (FIRST NAME ONLY): Aiping
PRESENTER (LAST NAME ONLY): Chen
PRESENTER (INSTITUTION ONLY): Los Alamos National Lab
PRESENTER (EMAIL ONLY): apchen@lanl.gov
PRESENTER (COUNTRY ONLY): United States
DATE/TIME SUBMITTED: September 25, 2023, 07:13 PM
ABSTRACT BODY:
Abstract Body: Current digital computing based on Von Neumann architecture suffers from several key bottleneck. Developing new computing models based on emergent devices provide solutions towards Beyond Moore's computing. Oxide-based memristive devices with excellent scalability have the potential to revolutionize not only the field of information storage but also neuromorphic computing. In this talk, I will first discuss some basics of the brain, brain-inspired neuromorphic computing and artificial intelligence. I will then discuss the roles of defects and interfaces on switching behavior in different types of memristive devices and their impacts on neuromorphic computing. This talk also shows how incompetence from the brain affects the understanding of emergent devices for applications.

CONTROL ID: 3975520
CURRENT SYMPOSIUM: S13 - Frontiers in Ferroic oxides: Synthesis, Properties, and Applications
CURRENT SESSION: Advanced synthesis and characterization techniques for ferroic oxides
TITLE: Anomalous Hall and Topological Hall Effects in $NiCo_2O_4$ Thin Films and Free-Standing Membranes
PRESENTATION TYPE: Invited (by Invitation Only)
PRESENTER (FIRST NAME ONLY): Xia
PRESENTER (LAST NAME ONLY): Hong
PRESENTER (INSTITUTION ONLY): University of Nebraska-Lincoln
PRESENTER (EMAIL ONLY): xia.hong@unl.edu
PRESENTER (COUNTRY ONLY): United States
DATE/TIME SUBMITTED: August 18, 2023, 04:08 PM
ABSTRACT BODY:
Abstract Body: The inverse spinel ferrimagnetic $NiCo_2O_4$ possesses high Curie temperature and spin polarization, making it a promising material candidate for spintronic applications. The magnetic state of $NiCo_2O_4$ films depends sensitively on the strain and disorder effects, which can lead to unconventional spin transport. In this talk, I will discuss our recent studies of anomalous Hall effect (AHE) and topological Hall effect (THE) in epitaxial $NiCo_2O_4$ thin films and free-standing $NiCo_2O_4$ membranes. $NiCo_2O_4$ films deposited on (001) $MgAl_2O_4$ substrates exhibit strong perpendicular magnetic anisotropy down to 1.2 nm thickness, while the (110) films possess in-plane anisotropy. The AHE exhibits a nonmonotonic temperature dependence and sign reversal driven by both temperature and film thickness, revealing the intricate interplay between the impurity scattering, band intrinsic Berry curvature, and correlation effect. THE has been observed in (110) $NiCo_2O_4$ films and (001) $NiCo_2O_4$ membranes at low temperature. The former has been correlated with the magnetic bubble domain formation, as revealed via magnetic force microscopy, suggesting the emergence of bimeron spin textures. The latter can be sensitively tuned by magnetic field cooling. Our study provides the material strategy for designing spin transport in $NiCo_2O_4$ via strain and disorder, paving the path for its technological implementation.

CONTROL ID: 3976652
CURRENT SYMPOSIUM: S12 - Magnetic and Superconducting Materials: From Basic Science to Applications
CURRENT SESSION: Other magnetic materials
TITLE: Vitrovac/PZT Magnetoelectric Laminate Composites for Wireless Power Transfer for In-Vivo Biomedical Devices
PRESENTATION TYPE: Contributed (Oral)
PRESENTER (FIRST NAME ONLY): Merissa
PRESENTER (LAST NAME ONLY): Hall
PRESENTER (INSTITUTION ONLY): United States Naval Academy
PRESENTER (EMAIL ONLY): hallmerissa@gmail.com
PRESENTER (COUNTRY ONLY): United States
DATE/TIME SUBMITTED: September 19, 2023, 07:33 PM
ABSTRACT BODY:
Abstract Body: The Internet of Things (IoT) has offered opportunities in modern medicine for the development of in-vivo biomedical devices that can perform monitoring and treatment for patients. However, these implanted devices require power, often coming from external wiring or implanted batteries. The use of external wires and batteries result in patient discomfort due to longer hospital stays, increased infection risks, and periodic surgeries. As a result, there is a need for non-invasive methods to remotely power implanted devices. One solution is the use of magnetoelectric materials, which generate an ac voltage when ac and dc magnetic fields are applied, resulting from the interaction of magnetostrictive and piezoelectric materials. Composite laminates are one class of magnetoelectric material, which consist of magnetostrictive and piezoelectric films physically stacked and bonded together. This work aims to fabricate and characterize a laminate composite of magnetostrictive Vitrovac® 7600 T70 and piezoelectric lead zirconate titanate (PZT), and demonstrate its capability to power an in-vivo sensor requiring at least 10mW. Future work seeks to explore how the Vitrovac/PZT composite's magnetoelectric response changes with the addition of biocompatible coatings such as parylene.

CONTROL ID: 3976670
CURRENT SYMPOSIUM: S14 - Emerging Semiconductor Materials and Interfaces
CURRENT SESSION: Other
TITLE: Identifying long-carrier lifetime semiconductors for photovoltaic and infra-red applications
PRESENTATION TYPE: Invited (by Invitation Only)
PRESENTER (FIRST NAME ONLY): Geoffroy
PRESENTER (LAST NAME ONLY): Hautier
PRESENTER (INSTITUTION ONLY): Dartmouth College
PRESENTER (EMAIL ONLY): Geoffroy.T.F.Hautier@dartmouth.edu
PRESENTER (COUNTRY ONLY): United States
DATE/TIME SUBMITTED: August 24, 2023, 10:37 AM
ABSTRACT BODY:
Abstract Body: Non-radiative lifetime controls the efficiency of many devices from solar cells to infra-red detectors. The Shockley-Read-Hall (SRH) processes where point defect help the carrier recombination have been especially singled-out as responsible for controlling lifetime in semiconductors. In the photovoltaic field, this created the push towards the search for materials exhibiting "defect-tolerance".
First principles computations can shine light into defect-induced recombination by providing defect formation energies, charge transition levels and even capture coefficients.
I will report on our efforts in this field of defect computations highlighting some of our work on earth-abundant photovoltaic materials (e.g., Zn_3P_2) but also on well-established materials used in infra-red detectors such as MCT ($Hg_{1-x}Cd_xTe$). Next, I will focus on how we are bringing these defects computations to computational screening with the search for new solar absorbers with exceptional carrier lifetime and defect-tolerance. Screening thousands of materials including their defect behavior, I will show how a handful materials exhibit favorable defects making them excellent candidates for opto-electronic applications and solar cells. I will especially highlight a family of materials for which recent experimental results confirm our computational predictions.

CONTROL ID: 3976969

CURRENT SYMPOSIUM: S8 - Data-driven and model-supported structure-property relationships in complex electroceramics

CURRENT SESSION: Local structure and properties of complex electroceramics

TITLE: Processing-composition-structure effects on the optical band-gap of ferroelectric $KNbO_3$-based ceramics

PRESENTATION TYPE: Invited (by Invitation Only)

PRESENTER (FIRST NAME ONLY): Cristina

PRESENTER (LAST NAME ONLY): Pascual-Gonzalez

PRESENTER (INSTITUTION ONLY): ICMM-CSIC

PRESENTER (EMAIL ONLY): cristina.pascual@csic.es

PRESENTER (COUNTRY ONLY): Spain

DATE/TIME SUBMITTED: September 25, 2023, 04:14 AM

ABSTRACT BODY:

Abstract Body: This work reviews and presents a study of band-gap engineering of ferroelectric solid-solutions based on $KNbO_3$. The strategy to narrow the band-gap of the parent $KNbO_3$ (3.22 eV), relies on replacing Nb^{5+} by lower valence transition metals (Me^{3+}) and K^{+1} by cations which maintain the compositions stoichiometric. All solid-solutions (1-x) $KNbO_3$-x $BiMeO_3$ (Me= Mn, Co and Ni) systems in a compositional range of 0≤x≤0.25 evolve from orthorhombic (x=0) to pseudo-cubic symmetry with an increase of x. Indeed, these two symmetries seem to coexist for intermediary concentrations. A solubility limit for orthorhombic $KNbO_3$ phase is determined for each system. In addition, a continuous band-gap narrowing was observed in all systems. Nevertheless, (1-x) $KNbO_3$-x $BiFeO_3$ (0≤x≤0.25) system maintained the polar phase up to x=0.25 and its band-gap was narrowed down to 2.22 eV. A photocurrent of 0.24 μA/cm^2 was measured for 0.75 $KNbO_3$- 0.25 $BiFeO_3$.

CONTROL ID: 3977001
CURRENT SYMPOSIUM: S18 - Structure and function in high-entropy, entropy-stabilized, and related oxides
CURRENT SESSION: Characterization and quantification of structure and properties
TITLE: Memristor Function From Tunable Carrier Transport and Dynamics in Entropy-Stabilized Oxides
PRESENTATION TYPE: Invited (by Invitation Only)
PRESENTER (FIRST NAME ONLY): John
PRESENTER (LAST NAME ONLY): Heron
PRESENTER (INSTITUTION ONLY): University of Michigan
PRESENTER (EMAIL ONLY): jtheron@umich.edu
PRESENTER (COUNTRY ONLY): United States
DATE/TIME SUBMITTED: August 26, 2023, 01:51 PM
ABSTRACT BODY:
Abstract Body: Entropy-stabilized oxides demonstrate a new degree of chemical control in materials and the opportunity to explore the interplay between local chemical and structural disorder and physical properties. Here, I will discuss the control of stereochemically-driven structural disorder in single crystalline, rocksalt, (MgCoNiCuZn)O-type entropy-stabilized oxides through the incorporation of Co, Cu, and Mg cations. Point defect formation energies from first principles calculations predict the local strain is the driver for defect formation and largely dictated by Cu incorporation. Electron transport measurements reveal a transition from Schottky to hopping conduction. The hopping conductivity is tunable via the Mg concentration (inversely correlated to Cu). Pulse measurements reveal a low resistance state with the short retention time that can be harnessed for memristor function. As theory predicts filled defect levels deep within the bandgap, we interpret the high resistance state as Schottky emission via these deep level states and the low resistance state at high applied voltage as electron hopping conduction through empty deep level states via the excitation of electrons from the defect levels. Our findings demonstrate novel interplay between structure and charge transport in entropy-stabilized oxides for tunable electronic responses and transformative computing applications.

CONTROL ID: 3977184
CURRENT SYMPOSIUM: Plenary Session
CURRENT SESSION: None
TITLE: Adsorption, Diffusion, and Disconnection Motion During Sintering: Engineering Microstructures by Grain Boundary Control
PRESENTATION TYPE: Plenary
PRESENTER (FIRST NAME ONLY): Wayne
PRESENTER (LAST NAME ONLY): Kaplan
PRESENTER (INSTITUTION ONLY): Technion - Israel Institute of Technology
PRESENTER (EMAIL ONLY): kaplan@technion.ac.il
PRESENTER (COUNTRY ONLY): Israel
DATE/TIME SUBMITTED: August 28, 2023, 11:01 AM
ABSTRACT BODY:
Abstract Body: At the temperatures required for densification, grain boundary (GB) motion often occurs, resulting in grain growth driven by the reduction of GB area. GB motion can result in pore detachment forming occluded porosity, requiring bulk diffusion for densification. Grain growth is often attributed to diffusional processes similar to those occurring during densification. However, recent evidence suggests that the mechanism of GB motion is via the motion of GB disconnections, which do not require long range diffusional transport.

Alumina is a paradigm for microstructural evolution studies. Mg is known to promote densification and limit grain growth by solute-drag. Ca causes exaggerated grain growth in alumina, and Ca in alumina at concentrations below the solubility limit increases GB mobility (solute-acceleration). The mechanism for solute-drag and solute-acceleration is believed to be adsorbing dopants (and their charge compensating defects) interacting with disconnections at GBs. Dopant adsorption to the surfaces of pores can change their relative mobility, where Mg is known to increase surface diffusion (preventing pore detachment). Cr apparently reduces surface diffusivity, increasing the probability for pore detachment and retarding densification. This lecture will review these concepts and define open questions for future research.

CONTROL ID: 3977825
CURRENT SYMPOSIUM: S15 - Electronics Materials for 5G/6G Applications
CURRENT SESSION: Materials growth and design
TITLE: Superior Microwave Properties and Structural Analysis of Alumina-based Ceramics
PRESENTATION TYPE: Contributed (Oral)
PRESENTER (FIRST NAME ONLY): Daniel
PRESENTER (LAST NAME ONLY): Tan
PRESENTER (INSTITUTION ONLY): Guangdong Technion-Israel Institute of Technology
PRESENTER (EMAIL ONLY): daniel.tan@gtiit.edu.cn
PRESENTER (COUNTRY ONLY): China
DATE/TIME SUBMITTED: September 05, 2023, 11:53 PM
ABSTRACT BODY:
Abstract Body: Al_2O_3 is an outstanding candidate for millimeter-wave dielectric ceramic devices due to its ultrahigh quality factor (Qf ~ 333,000 GHz) among known low permittivity (ε_r ~ 9.6) ceramics, however it bears with negative temperature coefficient of resonant frequency (TCF ~ -60 ppm/C). Although earlier literature studied the effect of TiO_2, annealing, and particle size on balancing TCF for its positive value, the TiO_2 concentration is higher than 8 mol% and higher Qf is yet to be achieved. Through a systematical study of sintering behaviors, subsequent heat treatments, and phase analysis of Al_2O_3 - x wt%TiO_2, the authors can achieve the superior overall properties with near zero TCF following subsequent annealing at lower temperatures. The best results obtained so far are attributed to the elimination of the secondary phase Al_2TiO_5 through decomposing into TiO_2 and their redistribution in the ceramic body.

The advantage of using submicron alumina particles is also demonstrated in terms of the density and Qf value. 150-nm alpha particles are found to be superior to 2-μm particles and 30-nm gamma alumina particles. The microstructure influence on properties and the root causes were analyzed using XRD, SEM/mapping, DSC, and dielectric measurements.

CONTROL ID: 3977948
CURRENT SYMPOSIUM: S12 - Magnetic and Superconducting Materials: From Basic Science to Applications
CURRENT SESSION: Other magnetic materials
TITLE: Magnetism by design: Validating computational predictions of Heusler skyrmion hosts
PRESENTATION TYPE: Contributed (Oral)
PRESENTER (FIRST NAME ONLY): Eve
PRESENTER (LAST NAME ONLY): Mozur
PRESENTER (INSTITUTION ONLY): Colorado School of Mines
PRESENTER (EMAIL ONLY): evemozur@mines.edu
PRESENTER (COUNTRY ONLY): United States
DATE/TIME SUBMITTED: August 31, 2023, 06:06 PM
ABSTRACT BODY:
Abstract Body: Developing control over large, non-collinear magnetic structures is essential for next generation magnetic technologies, including spintronics. However, the formation of large magnetic bodies requires a delicate balance of magnetic coupling strengths and spin-orbit interactions, such as the Dzyaloshinskii–Moriya interaction and the uniaxial magnetic anisotropy. Computational predictions suggest that these interactions can be controlled through compositional tuning. Here, we compare experimental results to the computational predictions for a subclass of intermetallic Heusler materials.

CONTROL ID: 3978733
CURRENT SYMPOSIUM: S13 - Frontiers in Ferroic oxides: Synthesis, Properties, and Applications
CURRENT SESSION: Advanced synthesis and characterization techniques for ferroic oxides
TITLE: Design, synthesis, and symmetry-dependent properties of ferroaxial materials
PRESENTATION TYPE: Invited (by Invitation Only)
PRESENTER (FIRST NAME ONLY): Tsuyoshi
PRESENTER (LAST NAME ONLY): Kimura
PRESENTER (INSTITUTION ONLY): University of Tokyo
PRESENTER (EMAIL ONLY): tkimura@ap.t.u-tokyo.ac.jp
PRESENTER (COUNTRY ONLY): Japan
DATE/TIME SUBMITTED: September 05, 2023, 02:40 AM
ABSTRACT BODY:
Abstract Body: The spontaneous symmetry breakdown of the matter is one of the most important concepts in materials physics and leads to a phase transition into an ordered phase and domain formation in its consequence. The so-called "ferroaxial order" characterized by a rotational structural distortion with an axial vector symmetry has gained growing interest as a new class of ordered state. However, there have been no systematic studies on the search for materials showing ferroaxial order. Only a few ferroaxial materials which show a pure ferroaxial transition were reported to date [e.g., $RbFe(MoO_4)_2$, $NiTiO_3$].
In this presentation, we report on the design, synthesis, and symmetry-dependent properties of ferroaxial materials. We propose that the staggard structure is crucial for realizing ferroaxial order in crystalline materials. We also propose a way to explore ferroaxial materials and conducted the formula-based screening using a regular expression search from a materials database and extracted ferroaxial candidates. As symmetry-dependent properties of ferroaxial materials, we show experimental demonstrations of electric field-induced optical rotation (i.e., electrogyration) and electric field-induced magnetochiral dichroism in a ferroaxial material, $NiTiO_3$.

CONTROL ID: 3978801
CURRENT SYMPOSIUM: S1: Characterization of Structure-Property Relationships in Functional Ceramics
CURRENT SESSION: Advances in connecting local and global structure to properties
TITLE: Doped $BiFe_{0.95}Co_{0.05}O_3$ Nanoparticles for Piezophotocatalysis
PRESENTATION TYPE: Contributed (Oral)
PRESENTER (FIRST NAME ONLY): Astita
PRESENTER (LAST NAME ONLY): Dubey
PRESENTER (INSTITUTION ONLY): University of Duisburg Essen
PRESENTER (EMAIL ONLY): astita.dubey@uni-due.de
PRESENTER (COUNTRY ONLY): Germany
DATE/TIME SUBMITTED: September 05, 2023, 07:41 AM
ABSTRACT BODY:
Abstract Body: Visible light absorber Bismuth Ferrite ($BiFeO_3$; BFO) nanoparticles (NPs) are potential materials in the field of optoelectronics, and catalysis due to their narrow band gap (2.18 eV) and robustness towards humidity and light. Nevertheless, the photocatalytic activity of BFO NPs is far behind the well-established TiO_2 because of faster recombination of charge carriers and formation of defect assisted trap states. Ionic engineering is one of the tools to tailor the photocatalytic activity. We probed the effect of Bi site doping into $BiFe_{0.95}Co_{0.05}O_3$ (BFC) NPs on the photodegradation of organic pollutants. The Bi-site doped BFC NPs are highly crystalline and single phase NPs. The piezoelectricity of Bi-site doped BFC NPs is studied and utilized for further piezo-photocatalysis.
The piezo-photocatalytic activity of these ferroelectric photocatalysts is studied under ultrasonication at pH 2. It is found that several types of dopants at the Bi-site alter the piezo-photocatalytic activity of the ferroelectric BFC NPs. The correlation of the photocatalytic activity with structural, optical, and electrical properties of the doped BFC NPs is established. The increased Piezoresponse due to Bi-site doping is one of the major factors in enhancing the photocatalytic performance of BFC NPs along with stronger light absorption in the visible range, larger surface area of NPs, and new active sites.

CONTROL ID: 3978845
CURRENT SYMPOSIUM: S7 - Defects and Transport in Ceramics
CURRENT SESSION: Defect-mediated properties
TITLE: Engineering surface dipoles on mixed conducting oxides with ultra-thin oxide decoration layers
PRESENTATION TYPE: Contributed (Oral)
PRESENTER (FIRST NAME ONLY): Matthaeus
PRESENTER (LAST NAME ONLY): Siebenhofer
PRESENTER (INSTITUTION ONLY): Massachusetts Institute of Technology
PRESENTER (EMAIL ONLY): msieben@mit.edu
PRESENTER (COUNTRY ONLY): United States
DATE/TIME SUBMITTED: September 05, 2023, 10:23 AM
ABSTRACT BODY:
Abstract Body: The surfaces of mixed ionic and electronic conducting (MIEC) materials are vital for many energy and sensing application, such as solid oxide cells, catalysts, and oxygen permeation membranes. Enhancing their properties is thus an active research area. Sub-monolayer oxide modifications have recently shown promising potential in altering several properties of MIEC surfaces, such as the work function and the catalytic activity toward the oxygen exchange reaction (OER). However, underlying mechanisms and the roles of specific charge carriers and defects are still areas of conjecture.

In this contribution, we investigated the effects of ultra-thin oxide layers on the properties of uncontaminated MIEC oxide surfaces combining in-situ impedance spectroscopy during pulsed laser deposition (i-PLD), near ambient pressure X-ray photoelectron spectroscopy (NAP-XPS) and density functional theory (DFT). Our study demonstrates a systematic modulation of MIEC oxide surfaces by decorating with oxides of varying ionic potential or acidity. As fundamental underlying mechanism we could identify charge redistribution processes and dipole formation at the heterojunction between host material and decoration. We explored the effect of surface decorations on the OER kinetics of various solid oxide fuel cell cathode materials, highlighting their potential to fine-tune reaction rates on MIEC surfaces.

CONTROL ID: 3978857
CURRENT SYMPOSIUM: S4 - Complex Oxide Thin Films and Heterostructures: From Synthesis to Strain/Interface-engineered Emergent Properties
CURRENT SESSION: Characterizations of strain, defects, and interfaces
TITLE: Investigating emerging interfaces by in-situ impedance spectroscopy during pulsed laser deposition
PRESENTATION TYPE: Contributed (Oral)
PRESENTER (FIRST NAME ONLY): Matthaeus
PRESENTER (LAST NAME ONLY): Siebenhofer
PRESENTER (INSTITUTION ONLY): Massachusetts Institute of Technology
PRESENTER (EMAIL ONLY): msieben@mit.edu
PRESENTER (COUNTRY ONLY): United States
DATE/TIME SUBMITTED: September 11, 2023, 09:28 AM
ABSTRACT BODY:
Abstract Body: In many devices and technologies, such as batteries, solid oxide cells, or sensors, interfaces are the critical locations where decisive reactions and transport processes occur. Modification of these interfaces is hence a widely applied method to improve device performance, well known from stabilizing coatings in battery materials or from surface modifications to enhance oxygen exchange in solid oxide fuel and electrolysis cells.

In this contribution we present in-situ impedance spectroscopy during pulsed laser deposition as a means of engineering and simultaneously investigating the electrochemical properties of emerging interfaces and heterostructures. We investigated the effect of oxidic decorations on the oxygen exchange kinetics and the sub-surface conductivity of mixed conducting oxides such as $La_{0.6}Sr_{0.4}CoO_{3-\delta}$ or $Pr_{0.1}Ce_{0.9}O_{2-\delta}$. We found a substantial alteration of reaction kinetics while the conductivity remains unaffected. Complemented by X-ray photoelectron spectroscopy and density functional theory, we could identify dipole formation at the heterojunction between the two oxides as the fundamental underlying process. Ultimately, we extend this method towards the electrochemical investigation of growing $La_{0.6}Sr_{0.4}CoO_3/SrTi_{0.3}Fe_{0.7}O_3$ multilayer structures, highlighting the extensive characterization capabilities of this technique.

CONTROL ID: 3979641

CURRENT SYMPOSIUM: S5 - Controlling Grain Boundary Structure, Chemistry, and their Network as a Function of Mtl Processing

CURRENT SESSION: Atomic structure, chemistry, bonding configuration

TITLE: Probing the vibrations of individual dislocation cores in low-angle grain boundaries using monochromated electron energy-loss spectroscopy

PRESENTATION TYPE: Invited (by Invitation Only)

PRESENTER (FIRST NAME ONLY): Jordan

PRESENTER (LAST NAME ONLY): Hachtel

PRESENTER (INSTITUTION ONLY): Oak Ridge National Laboratory

PRESENTER (EMAIL ONLY): Jordan.Hachtel@gmail.com

PRESENTER (COUNTRY ONLY): United States

DATE/TIME SUBMITTED: September 21, 2023, 06:03 PM

ABSTRACT BODY:

Abstract Body: Aberration-corrected scanning transmission electron microscopy (AC-STEM) has been an extremely impactful tool in the study of grain boundaries due to its ability to extract atomic-resolution structural and compositional information using an Ångstrom scale probe with electron energy-loss spectroscopy (EELS). The energy resolution of EELS on most AC-STEMs because the energy-resolution of the instrument is between 0.3-0.8 eV, which while perfectly satisfactory for 'core-loss' excitaitons (> 50 eV) can be challenging for 'low-loss' excitations (1-5 eV) and completely prohibitive for the vibrational regime (< 0.5 eV). However, the energy resolution can be improved significantly using monochromated AC-STEM (MAC-STEM), which can now obtain energy resolution down to a level of 0.002-0.003 eV, without significantly impacting the spatial resolution.

Here, I will show applications of MAC-STEM to complex oxides, and specifically to a symmetric 10° $SrTiO_3$ grain boundary, where we obtain sufficient simultaneous spatial and energy resolution to probe localized vibrational modes inside individual dislocation cores of the grain boundary. I will also explain the experimental considerations needed to probe localized vibrational modes with MAC-STEM, and how to utilize MAC-STEM to unveil emergent thermal behavior in oxide heterostructures.

CONTROL ID: 3979971
CURRENT SYMPOSIUM: S2 - Advanced Microelectronics
CURRENT SESSION: Role of defects, interface and synthesis/fabrication process on device properties
TITLE: Self-assembled Vertical Nanopillars in Oxide Thin Films: A Material Platform for Giant Piezoelectricity and Synaptic Electronics
PRESENTATION TYPE: Invited (by Invitation Only)
PRESENTER (FIRST NAME ONLY): Huajun
PRESENTER (LAST NAME ONLY): Liu
PRESENTER (INSTITUTION ONLY): Institute of Materials Research and Engineering, A*STAR, Singapore
PRESENTER (EMAIL ONLY): liu_huajun@imre.a-star.edu.sg
PRESENTER (COUNTRY ONLY): Singapore
DATE/TIME SUBMITTED: September 22, 2023, 01:28 AM
ABSTRACT BODY:
Abstract Body: Transition metal oxide thin films exhibit many interesting properties, including ferroelectricity, piezoelectricity, superconductivity and electrochemical activities for energy applications. Nanocomposite oxide thin films provide additional freedom to integrate multiple functionalities into one material platform. In this talk, I will discuss our recent work on self-assembled vertical nanopillars in $NaNbO_3$ (NNO) epitaxial thin films. Through precise control of Na deficiency, Nb atoms occupy vacant Na positions and form the Nb_{Na} antisite defects in NNO films. These antisite defects self-assembled to form vertical nanopillar structures. Such nanopillar regions with local structural and polar heterogeneity give rise to giant piezoelectric coefficients while maintain a high Curie temperature (Science, 369, 292, 2020). In addition, the self-assembled vertical nanopillars also give rise to resistive switching behaviour and tuneable resistance states by electric field. The NNO film based devices exhibit synaptic plasticity under applied electrical pulses, promising for the application in neuromorphic computing.

CONTROL ID: 3980116
CURRENT SYMPOSIUM: S10 - Fundamental mechanisms in materials for next generation computing devices
CURRENT SESSION: Phase transitions in 2D materials and devices
TITLE: Large second harmonic generation in 2D metal thiophosphates
PRESENTATION TYPE: Invited (by Invitation Only)
PRESENTER (FIRST NAME ONLY): Michael
PRESENTER (LAST NAME ONLY): Susner
PRESENTER (INSTITUTION ONLY): Air Force Research Laboratory
PRESENTER (EMAIL ONLY): michael.susner.1@us.af.mil
PRESENTER (COUNTRY ONLY): United States
DATE/TIME SUBMITTED: September 11, 2023, 10:51 AM
ABSTRACT BODY:
Abstract Body: Nonlinear optics are major component of modern laser systems and current and next-generation optoelectronic devices. The dominant SHG interaction mechanism involves light-matter interactions in materials lacking inversion symmetry thus making ferfoelectric materials a particular and fruitful target of investigation. In addition, 2D materials allow for new opportunities in next-generation data storage, communication, spintronic computing, nanoelectronics, and sensors due, in part, to the lower barriers within their potential energy landscapes that render them more susceptible to and their clean, exfoliatable surfaces which in turn allow for novel functionalities through stacking and twisting to create heterostructures. We have begun a program of study uniting these two concepts through our investigation of 2D ferroelectric compounds in the metal thiophosphate family. In particular, we have studied the 2-photon response and second harmonic generation properties of $CuInP_2S_6$, $AgScP_2S_6$, and $CuScP_2S_6$. These materials show diverse properties, including large two-photon absorption, low optical limiting response, and high laser damage threshold.

CONTROL ID: 3980127
CURRENT SYMPOSIUM: S12 - Magnetic and Superconducting Materials: From Basic Science to Applications
CURRENT SESSION: Other magnetic materials
TITLE: Investigation of magnetic 2D Multiferroic metal thiophosphate materials
PRESENTATION TYPE: Invited (by Invitation Only)
PRESENTER (FIRST NAME ONLY): Michael
PRESENTER (LAST NAME ONLY): Susner
PRESENTER (INSTITUTION ONLY): Air Force Research Laboratory
PRESENTER (EMAIL ONLY): michael.susner.1@us.af.mil
PRESENTER (COUNTRY ONLY): United States
DATE/TIME SUBMITTED: September 11, 2023, 11:04 AM
ABSTRACT BODY:
Abstract Body: Exotic 2D materials with ferroic properties such as magnetism and ferroelectricity are the current subject of intense investigation for integration into new device paradigms for next-generation electronics. In this presentation, we will discuss three different members of the metal thiophosphate family and the complex relationships between their structural and magnetic properties. First, we present an overview of magnetic MTP materials which details their potential as a powerful materials platform with promise for nanoscale electronic and optoelectronic devices. We will then detail our investigations into the multiferroic compound 2D material $CuCrP_2S_6$ in terms of a detailed analysis of its antiferromagnetic and incommensurate antiferroelectric phase transitions. We will then discuss the magnetic, calorimetric, and structural behavior of AgCrP2Se6, which exhibits an antiferromagnetic transition at 42 K with spins pointed along the lamellar direction (i.e. the ab plane) rather than along the stacking direction. We note the presence of additional indicators of structural rearrangements at higher temperatures from calorimetric, powder diffraction, and Raman analyses which suggest more complex and possible multiferroic behavior.

CONTROL ID: 3980193

CURRENT SYMPOSIUM: S3 - Advanced Electronic Materials: Processing and Structures and Properties and Applications

CURRENT SESSION: Materials design, new materials and structures, and their emerging applications

TITLE: Phase Boundaries of $LaWN_3$ Synthesized from Oxide Precursors via Ammonolysis

PRESENTATION TYPE: Contributed (Oral)

PRESENTER (FIRST NAME ONLY): Shannon

PRESENTER (LAST NAME ONLY): Rogers

PRESENTER (INSTITUTION ONLY): Colorado School of Mines

PRESENTER (EMAIL ONLY): shannonrogers@mines.edu

PRESENTER (COUNTRY ONLY): United States

DATE/TIME SUBMITTED: October 09, 2023, 12:27 PM

ABSTRACT BODY:

Abstract Body: Many nitride perovskites have been computationally determined to be stable, yet few of these materials have been synthesized in bulk form. The nitride material discovery community has devoted significant efforts to developing nitride ceramic thin films, including the first nitride perovskite, $LaWN_3$, which shows promising piezoelectric response. Here, we synthesize $LaWN_3$ and oxynitride ($LaW(O,N)_3$) bulk powders from oxide precursor $La_2W_2O_9$ by ammonolysis with varying La:W to determine phase space and defect accommodation present in this complex oxynitride system. These results establish a foundation for studying the effects of vacancies and other point defects during bulk nitride sintering. The $LaWN_3$ and $LaW(O,N)_3$ perovskite compounds have higher defect accommodation and are more tolerant to off-cation-stoichiometry compared to the $La_2W_2O_9$ precursor compound. We are investigating the roles that anion site occupancy and cation valence states play in solid solubility asymmetry.

CONTROL ID: 3980540
CURRENT SYMPOSIUM: S12 - Magnetic and Superconducting Materials: From Basic Science to Applications
CURRENT SESSION: Other magnetic materials
TITLE: Electrophoretic deposition of iron oxide nanoparticles in various planar geometries for on-chip component fabrication
PRESENTATION TYPE: Contributed (Oral)
PRESENTER (FIRST NAME ONLY): Sara
PRESENTER (LAST NAME ONLY): Mills
PRESENTER (INSTITUTION ONLY): US Naval Research Laboratory
PRESENTER (EMAIL ONLY): sara.mills.ctr@nrl.navy.mil
PRESENTER (COUNTRY ONLY): United States
DATE/TIME SUBMITTED: September 18, 2023, 12:14 PM
ABSTRACT BODY:
Abstract Body: The miniaturization of on-chip components for integrated circuits has become paramount as devices decrease in size and operating frequencies increase, all while demanding equivalent or improved performance. Many of these components are comprised of a magnetic material, which must be incorporated in a semiconductor compatible process with fine control over the deposition area. Electrophoretic deposition (EPD), a process in which charged particles are deposited onto a conductive substrate under an applied dc electric field, can deposit micron-thick films of magnetic nanoparticles into feature sizes around 1 µm. This work shows the development of EPD to deposit iron oxide nanoparticles into molds of various sizes and geometries patterned with photoresist on copper-cladded printed circuit board (PCB) substrates. Due to a difference between the dielectric constant of the photoresist and EPD suspension, the nanoparticles consolidated on the exposed regions of the substrate in molds with various geometric shapes and dimensions down to 1 µm. This work serves as the initial step for fabricating 0-3 nanocomposites as on-chip components, such as metaconductors. The iron oxide nanoparticle film, deposited in a well-defined area, could then be infiltrated to allow for the quick fabrication of 0-3 composite metaconductors with decreased resistance at high frequencies.

CONTROL ID: 3980720
CURRENT SYMPOSIUM: S13 - Frontiers in Ferroic oxides: Synthesis, Properties, and Applications
CURRENT SESSION: Advanced synthesis and characterization techniques for ferroic oxides
TITLE: Topotactic phase transitions in complex oxide membranes
PRESENTATION TYPE: Invited (by Invitation Only)
PRESENTER (FIRST NAME ONLY): Seung Sae
PRESENTER (LAST NAME ONLY): Hong
PRESENTER (INSTITUTION ONLY): University of California, Davis
PRESENTER (EMAIL ONLY): sshong@ucdavis.edu
PRESENTER (COUNTRY ONLY): United States
DATE/TIME SUBMITTED: September 12, 2023, 04:18 PM
ABSTRACT BODY:
Abstract Body: Topotactic phase transitions (TPTs) represent structural phase transitions involving oxidation/reduction processes, resulting in unusual atomic coordination and oxidation states of metal atoms. The dramatic structural phase transitions provide a unique switching mechanism between ferromagnetism, antiferromagnetism, and superconductivity. Recently, TPT has garnered substantial attention, owing to its implications in both fundamental science and technological applications from neuromorphic computing to unconventional superconductivity.

Despite remarkable advancements in the field, there has been a lack of research into the microscopic aspects of TPTs. Freestanding oxide membranes have advantages in studying the local evolution of TPTs at different length scales. This presentation will focus on atomic-scale microscopy studies of TPTs in strontium cobaltite ($SrCoO_{3-\delta}$) membranes based on the water-soluble layer etching technique. Whereas the macroscopic characterization showed clear structural and magnetic phase transitions within the membranes, transmission electron microscopy investigations show the intricate nature of these transitions inaccessible by other means. The observations demonstrate how freestanding oxide membranes can contribute to understanding the microstructure evolution during TPTs, as well as providing the guideline for the predictable switching phenomena in ferroic oxide materials.

CONTROL ID: 3981004
CURRENT SYMPOSIUM: S12 - Magnetic and Superconducting Materials: From Basic Science to Applications
CURRENT SESSION: Other magnetic materials
TITLE: 3D Printing of Magnetic Materials and Devices
PRESENTATION TYPE: Invited (by Invitation Only)
PRESENTER (FIRST NAME ONLY): Nathan
PRESENTER (LAST NAME ONLY): Lazarus
PRESENTER (INSTITUTION ONLY): University of Delaware
PRESENTER (EMAIL ONLY): nlazarus@udel.edu
PRESENTER (COUNTRY ONLY): United States
DATE/TIME SUBMITTED: September 15, 2023, 10:18 AM
ABSTRACT BODY:
Abstract Body: Additive manufacturing of electronics is a powerful approach for adding new functionality into 3D printed parts, allowing circuitry to be integrated directly into complex three dimensional structures. However, electromagnetic devices such as inductors and transformers have remained difficult to 3D print; printed conductors such as silver ink are both more resistive and less reliable than their bulk metal counterparts, and printable magnetic materials perform similarly poorly. I will talk about our work on using multimaterial 3D printing and hybrid processing techniques to create high performance electromagnetic devices. Electrochemical metallization is used to deposit copper and other metals selectively onto 3D printed parts, allowing bulk metal performance on printable conductors. This approach is demonstrated on a number of different 3D printable conductors including silver inks and conductive thermoplastic filaments. Ferrofluid, magnetic fluids consisting of nanoscale ferrites in a carrier liquid, are demonstrated to form magnetic cores in 3D printed magnetic devices without processing at high temperatures. Using this approach, we have demonstrated peak quality factors as high as 71 for a 3D printed coil.

CONTROL ID: 3981038

CURRENT SYMPOSIUM: S11 - In situ/operando Characterization of Nanomaterials

CURRENT SESSION: Advances in developing in situ characterization techniques related to electron microscopy, X-ray, and neutron characterization techniques

TITLE: Time-resolved neutron reflectometry study of lithium-mediated nitrogen reduction

PRESENTATION TYPE: Invited (by Invitation Only)

PRESENTER (FIRST NAME ONLY): Mathieu

PRESENTER (LAST NAME ONLY): Doucet

PRESENTER (INSTITUTION ONLY): Oak Ridge National Lab

PRESENTER (EMAIL ONLY): doucetm@ornl.gov

PRESENTER (COUNTRY ONLY): United States

DATE/TIME SUBMITTED: September 13, 2023, 10:11 AM

ABSTRACT BODY:

Abstract Body: Li-mediated electrochemical reduction of nitrogen to NH_3 has been identified as an alternative to the resource-demanding Haber-Bosch ammonia process. Through this process, an initial electrodeposition of Li is followed by a reaction to form Li_3N and subsequent ethanolysis of Li_3N to form NH_3.

We studied Li electrodeposition on a molybdenum thin film in contact with THF with 0.2 M $LiClO_4$ and 1 v% EtOH at a current density between -0.1 and -0.5 mA/cm^2. The neutron reflectometry data acquired during plating was sliced in 15- to 60-second intervals in a wide enough Q range to allow time-resolved modeling of the plating process. We will present our approach to time-resolved reflectometry and how it developed at the Liquids Reflectometer. We will also discuss the challenges involved in analyzing time-resolved data.

CONTROL ID: 3981050
CURRENT SYMPOSIUM: S15 - Electronics Materials for 5G/6G Applications
CURRENT SESSION: 5G/6G Devices
TITLE: Glass substrates in GHz electronics: How to provide reliable material data for accurate field simulations?
PRESENTATION TYPE: Invited (by Invitation Only)
PRESENTER (FIRST NAME ONLY): Martin
PRESENTER (LAST NAME ONLY): Letz
PRESENTER (INSTITUTION ONLY): Schott AG
PRESENTER (EMAIL ONLY): martin.letz@schott.com
PRESENTER (COUNTRY ONLY): Germany
DATE/TIME SUBMITTED: September 23, 2023, 04:49 PM
ABSTRACT BODY:
Abstract Body: Glasses get increasing attention for RF applications due to their high material homogeneity, their smooth surfaces, their relatively low dielectric constants and low dielectric loss in combination with small temperature coefficients. The ability to manufacture glass in panel sizes enables large size antenna arrays in a single package. Among glasses Borosilicate glasses have special RF properties. Boronoxide in glass strongly hinders alkali ion movement and therefore reduces dielectric losses. In addition low density glasses which corresponds to low dielectric constant (eps_0 ≈ 4) can be made with a high Boron content. In the present work we discuss the connection between glass composition and RF properties, give measurement details and show RF properties of different Borosilicate glasses for frequencies up to 110 GHz and give examples for RF devices which are made from such glasses.

CONTROL ID: 3981138
CURRENT SYMPOSIUM: S18 - Structure and function in high-entropy, entropy-stabilized, and related oxides
CURRENT SESSION: Characterization and quantification of structure and properties
TITLE: Enhanced Ionic Conductivity through Multication Doping of Ceria Oxides
PRESENTATION TYPE: Contributed (Oral)
PRESENTER (FIRST NAME ONLY): Elizabeth
PRESENTER (LAST NAME ONLY): Gager
PRESENTER (INSTITUTION ONLY): University of Florida
PRESENTER (EMAIL ONLY): egager@ufl.edu
PRESENTER (COUNTRY ONLY): United States
DATE/TIME SUBMITTED: September 25, 2023, 07:47 PM
ABSTRACT BODY:
Abstract Body: High entropy oxides (HEOs) have shown promise for tailoring the properties of materials through randomization of dopants. For ion-conductors, co-doped ceria oxides have shown superior ionic conductivity to singly-doped compounds, however, possible benefits of multication-doped ceria, and the origins of material differences, are not understood. Here, we present a study of the conductivity of a four-element doped ceria system, $Ce_{1-x}Pr_{x/4}Nd_{x/4}Sm_{x/4}Gd_{x/4}O_{2-\delta}$. To maintain homogeneity and stoichiometric control in the complex system, these materials were synthesized using the oxalate method. While the multication-doped materials have similar grain conductivity to singly-doped ceria, the grain boundary conductivity is increased by almost two orders of magnitude compared to conventional singly-doped ceria with an increase up to 15% doping (1.75 x 10^{-2} S cm^{-1} vs. 0.02 x 10^{-2} S cm^{-1} at 600 °C). We present a comprehensive analysis of crystallographic and elemental characterizations of the samples, including synchrotron pair distribution function (PDF) analysis. Structural changes resulting from compositional changes and processing are discussed in the context of the measured ionic conductivity. Our results suggest that dopant segregation, microstructure, and the resulting ionic conductivity of the grain boundary can be tailored with a multication doping approach.

CONTROL ID: 3981139
CURRENT SYMPOSIUM: S9 - Machine Learning and Automated Synthesis/Characterization for Novel Materials
CURRENT SESSION: Data-driven modeling and analysis
TITLE: Deep Learning Workflows for Microstructural Characterization of Materials
PRESENTATION TYPE: Contributed (Oral)
PRESENTER (FIRST NAME ONLY): Alexandre
PRESENTER (LAST NAME ONLY): Bordas
PRESENTER (INSTITUTION ONLY): University of Florida
PRESENTER (EMAIL ONLY): abordas@ufl.edu
PRESENTER (COUNTRY ONLY): United States
DATE/TIME SUBMITTED: September 25, 2023, 02:16 PM
ABSTRACT BODY:
Abstract Body: Microstructure analysis is essential in materials science and engineering and therefore, its accurate characterization is vital. However, it is typically time-consuming and requires manual input, which results in variability and inaccuracy. Here, we showcase two deep-learning end-to-end workflows for local and general image feature extraction that help overcome these limitations. First, for grain size determination, a pre-trained holistically nested edge detection model (HED) is used, achieving lower error (3.4%) over Canny's method (10.2%), benchmarked to Heyn's and planimetric methods. Second, for spinodal decomposition characterization, we apply transfer learning using a mask region convolutional neural network (R-CNN) on experimental and simulated (Allen-Cahn) microstructures. The result is an information-dense quantification of each micrograph that is exploited by machine learning models and thus enables more precise microstructural characterization. As a proof of concept, we quantify the spinodal decomposition marker, gamma, reducing computation time by at least one order of magnitude over traditional methods. Overall, the workflows enable faster and more consistent micrograph analysis over manual methods while providing complex feature quantification with ease. We will discuss avenues for adapting this approach to other materials characterization routines.

CONTROL ID: 3981191
CURRENT SYMPOSIUM: S2 - Advanced Microelectronics
CURRENT SESSION: Memristive switching and performance
TITLE: Memristive Nanodevices and Arrays for Unconventional Computing
PRESENTATION TYPE: Invited (by Invitation Only)
PRESENTER (FIRST NAME ONLY): Qiangfei
PRESENTER (LAST NAME ONLY): Xia
PRESENTER (INSTITUTION ONLY): University of Massachusetts Amherst
PRESENTER (EMAIL ONLY): qxia@umass.edu
PRESENTER (COUNTRY ONLY): United States
DATE/TIME SUBMITTED: September 13, 2023, 03:51 PM
ABSTRACT BODY:
Abstract Body: With the conductance state at each cell representing synaptic weights, memristive neural networks are capable of parallel in-memory computing using physical laws, bringing in orders of magnitude improvement in energy consumption and computing throughput.
We designed and fabricated a highly reliable Ta/HfO_2 device that meets most of the required properties for in-memory computing in artificial neural networks. The device exhibits multiple non-volatile and stable resistance levels that are tuned with low voltages in an analog fashion. The device also has a high switching speed and high endurance. We further integrated these devices into large analog crossbar arrays according to the one transistor one resistance switch (1T1R) architecture. The resistance of each level can be precisely programmed using electric pulses and peripheral feedback circuits for high-precision inference purposes. Linear and symmetric weight updating with minimal variation has also been demonstrated with a single shot programming scheme for online training. Finally, we implemented several neural network topographies such as single- and multiple-layer perceptron, long short-term memory, and convolutional neural networks. We demonstrated applications of these neural networks in image compression, object recognition, time-series regression, video processing, and reinforcement learning, to name a few.

CONTROL ID: 3981300

CURRENT SYMPOSIUM: S4 - Complex Oxide Thin Films and Heterostructures: From Synthesis to Strain/Interface-engineered Emergent Properties

CURRENT SESSION: Controlled synthesis of epitaxial thin films and nanocomposites, including heterostructures, superlattices, and vertically aligned nanocomposites for functional properties

TITLE: Antiferroelectric thin films with exotic polarization switching and abnormally high electromechanical response

PRESENTATION TYPE: Invited (by Invitation Only)

PRESENTER (FIRST NAME ONLY): Hao

PRESENTER (LAST NAME ONLY): Pan

PRESENTER (INSTITUTION ONLY): University of California, Berkeley

PRESENTER (EMAIL ONLY): hao.pan@berkeley.edu

PRESENTER (COUNTRY ONLY): United States

DATE/TIME SUBMITTED: September 14, 2023, 01:13 AM

ABSTRACT BODY:

Abstract Body: Antiferroelectrics (AFEs), which are characterized by antiparallel cation displacements and a field-induced reversible AFE-to-ferroelectric (FE) transition, exhibit great potential for dielectric, electrothermal and electromechanical applications. In this talk I will first demonstrate the epitaxial growth of high-quality AFE thin films (focusing on the prototype $PbZrO_3$). I will discuss how the point defects (introduced during the growth process) can effectively pin the AFE-FE phase transition boundaries and how this defect pinning effects can be used as a new freedom to regulate the polarization switching behaviors and adjust the dielectric tunability of the antiferroelectric films. I will then demonstrate an ultrahigh electromechanical response realized in judiciously designed AFE thin films with an unconventional coupling of the AFE-FE transition and the substrate constraint (clamping). While conventional FE materials, when fabricated in thin-film forms, exhibit severally degraded electromechanical responses due to substrate clamping, we reveal a positive impact of clamping on AFE films that adversely enhances the electromechanical responses and causes an abnormal thickness scaling of the performance. These works are expected to bring new insights into antiferroelectrics and to highlight the great opportunities of antiferroelectric films for various applications.

CONTROL ID: 3901341
CURRENT SYMPOSIUM: S3 - Advanced Electronic Materials: Processing and Structures and Properties and Applications
CURRENT SESSION: Advanced electronic materials, including ferroelectric, piezoelectric, dielectric, electrostrictive, electrocaloric and pyroelectric materials
TITLE: Antiferroelectric-like Tetragonal Tungsten Bronzes for Capacitor Applications
PRESENTATION TYPE: Contributed (Oral)
PRESENTER (FIRST NAME ONLY): Kerry
PRESENTER (LAST NAME ONLY): McMahon
PRESENTER (INSTITUTION ONLY): University of St Andrews
PRESENTER (EMAIL ONLY): kam28@st-andrews.ac.uk
PRESENTER (COUNTRY ONLY): United Kingdom
DATE/TIME SUBMITTED: September 22, 2023, 06:46 AM
ABSTRACT BODY:
Abstract Body: Multilayer ceramic capacitors (MLCCs) are globally prevalent in electronic devices (at >10^{12} units pa) and are predominantly based on the ferroelectric perovskite, barium titanate ($BaTiO_3$). However, due to its collapse in relative permittivity at both high temperature ($T>T_C$) and applied field, $BaTiO_3$ has limitations for applications under extreme conditions and meeting the demand for high-specification MLCCs is an increasing challenge. Antiferroelectrics provide a promising solution due to their enhanced high field and temperature-stable permittivity.

Tetragonal tungsten bronzes (TTBs) are a class of materials with the general formula $A1_2A2_4B1_2B2_8O_{30}$. They are structurally related to the commonly studied ABO_3 perovskites but with additional chemical and structural flexibility. Although ferroelectric TTBs have been widely studied, compositions which exhibit antiferroelectric-like behaviour have only recently been reported. Understanding the origins of the desired antiferroelectric-like behaviour is key for future development. The chemical and structural complexity of these materials, however, requires a suite of techniques including electron, neutron and synchrotron diffraction combined with dielectric, ferroelectric and second harmonic generation measurements to untangle the composition-structure-property interrelationships. Here we will discuss some of our recent work in this area.

CONTROL ID: 3981371

CURRENT SYMPOSIUM: S13 - Frontiers in Ferroic oxides: Synthesis, Properties, and Applications

CURRENT SESSION: Heavy transition metal oxides

TITLE: Towards the realization of Kitaev Quantum Spin Liquid State in 4d and 5d Honeycomb Magnets and beyond

PRESENTATION TYPE: Invited (by Invitation Only)

PRESENTER (FIRST NAME ONLY): Hidenori

PRESENTER (LAST NAME ONLY): Takagi

PRESENTER (INSTITUTION ONLY): Max Planck Institute for Solid State Research

PRESENTER (EMAIL ONLY): h.takagi@fkf.mpg.de

PRESENTER (COUNTRY ONLY): Germany

DATE/TIME SUBMITTED: September 16, 2023, 08:06 AM

ABSTRACT BODY:

Abstract Body: Alexei Kitaev proposed in 2006 a simple but elegant spin model on honeycomb lattice that is exactly solvable and has a quantum spin liquid (QSL) ground state, in which the spins fractionalize into exotic quasiparticles called Majorana fermions [1]. The proposal was followed by a theoretical recipe for the realization of the Kitaev model in materials using spin-orbit coupled J_{eff} = 1/2 pseudo-spins. It triggered an extensive search for the Kitaev QSL state and the Majorana fermions in a family of iridium (IV) and ruthenium (III) compounds where the d^5 transition metal ions octahedrally coordinated with anions form an edge-shared honeycomb network [2]. We are going to talk about the rapid progress, light and shadow, in the materialization the Kitaev QSL as well as the hunting of Majorana fermions through the detection of unusual thermal transport. The topics will include the QSL behavior and its relationship to the structural disorder of interlayer hydrogens in the 5d $H(D)_3LiIr_2O_6$ [3], the controversy of quantized thermal hall effect in the 4d $RuCl_3$ [4] and the extended materials exploration in 3d and 4f compounds [5].

[1] A. Kitaev, Ann. Phys. 321, 2 (2006).

[2] H. Takagi et al., Nat. Rev. Phys. 1, 264 (2019).

[3] K. Kitagawa et al., Nature 559, 227 (2018).

[4] J. Bruin et al, Nature Phys. 18, 401 (2022). APL Mater.10, 090703 (2022).

[5] Y. Matsumoto et al., arXiv:2207.02329.

CONTROL ID: 3981463
CURRENT SYMPOSIUM: S17 - Electronic and Ionic Materials in Battery Energy Storage
CURRENT SESSION: Solid State Ion Conductors
TITLE: Electronic Transport in Lithium Solid Electrolytes
PRESENTATION TYPE: Invited (by Invitation Only)
PRESENTER (FIRST NAME ONLY): Fudong
PRESENTER (LAST NAME ONLY): Han
PRESENTER (INSTITUTION ONLY): Rensselaer Polytechnic Institute
PRESENTER (EMAIL ONLY): hanf2@rpi.edu
PRESENTER (COUNTRY ONLY): United States
DATE/TIME SUBMITTED: September 14, 2023, 12:32 PM
ABSTRACT BODY:
Abstract Body: While significant efforts have been devoted to improving the ionic conductivity of lithium solid electrolytes (SEs), electronic transport, which has an important role in the calendar life, energy density, and cycling stability of solid-state batteries (SSBs), is rarely studied. In this talk, we will present our recent results on the electronic conductivities of three representative SEs, including Li3PS4, Li7La3Zr2O12, and Li3YCl6 for sulfide-, oxide-, and halide-based SEs respectively. By revisiting direct current polarizations using two-blocking-electrode cells and the Hebb-Wagner approach, we propose that the electronic conductivities of SEs are overestimated from the conventional measurements. We will discuss the limitations and sources of inaccuracy for these measurements and highlight the anodic decomposition of SE as the key source for the overestimated result. Modifications in the electrode selection and data interpretation are also proposed to approach the intrinsic electronic conductivity of SEs. To estimate the electronic conductivity of sulfides that decompose during measurement, a two-step polarization method will also be introduced. The electronic conductivities of all SEs measured by the modified approach are one or two orders of magnitude lower than the reported value. Despite that, the electronic conductivity of sulfides seems to be still quite high to enable SSBs with a long calendar life of >10 years.

CONTROL ID: 3981488
CURRENT SYMPOSIUM: S4 - Complex Oxide Thin Films and Heterostructures: From Synthesis to Strain/Interface-engineered Emergent Properties
CURRENT SESSION: Characterizations of strain, defects, and interfaces
TITLE: Mapping Strain and Polarization at Defects and Interfaces with Nanobeam Electron Diffraction
PRESENTATION TYPE: Invited (by Invitation Only)
PRESENTER (FIRST NAME ONLY): Megan
PRESENTER (LAST NAME ONLY): Holtz
PRESENTER (INSTITUTION ONLY): Colorado School of Mines
PRESENTER (EMAIL ONLY): meh282@cornell.edu
PRESENTER (COUNTRY ONLY): United States
DATE/TIME SUBMITTED: September 14, 2023, 02:13 PM
ABSTRACT BODY:
Abstract Body: Understanding strain and local polar ordering is key to understanding ferroelectricity in thin film systems, especially for systems with small domains, interfaces, or those with significant disorder. Scanning nanobeam electron diffraction (NBED) combined with new high speed, pixelated scanning transmission electron microscopy (STEM) detectors make it possible to measure a diffraction pattern (k_x, k_y) at every scan position (x, y). This opens doors to investigate lattice parameters, local fields, polarization directions, and charge densities with relatively low beam dose over large fields of view. Here we will discuss methods to extract picometer precise information on lattice strain and polarizations at the nanometer scale. Quantitatively extracting both the magnitude and direction of polarization vectors is still challenging, so here we will describe a cepstral approach, similar to a 2D pair-correlation function, to measure local polar displacements that drive ferroelectricity.

CONTROL ID: 3981507

CURRENT SYMPOSIUM: S11 - In situ/operando Characterization of Nanomaterials

CURRENT SESSION: Advances in developing in situ characterization techniques related to electron microscopy, X-ray, and neutron characterization techniques

TITLE: Structure and Dynamics in Electrolyte-gated $NdNiO_3$

PRESENTATION TYPE: Invited (by Invitation Only)

PRESENTER (FIRST NAME ONLY): Hao

PRESENTER (LAST NAME ONLY): Zheng

PRESENTER (INSTITUTION ONLY): Argonne National Lab

PRESENTER (EMAIL ONLY): hao.zheng@anl.gov

PRESENTER (COUNTRY ONLY): United States

DATE/TIME SUBMITTED: September 24, 2023, 11:06 PM

ABSTRACT BODY:

Abstract Body: In transition metal oxide, competition between comparable interactions on highly correlated electrons gives rise to ground states highly susceptible to lattice perturbations. One way of tuning the ground state is via an oxygen reduction process, where slight oxygen vacancies distort the lattice locally and effectively change their electrical, orbital, and magnetic properties. A good example is the rare-earth nickelate perovskites NdNiO3, the presence of oxygen deficiency suppresses the metal-insulator transition and drives the system into a metallic state. More interestingly, the electrostatic gating on NdNiO3 epitaxial thin film via ionic liquid has displayed a gigantic and reversible change of resistance with hysteresis, making it a potential candidate for data storage and neuromorphic computing. While it is believed that the formation of oxygen vacancy is essential to its electrical transport behavior, how the oxygen-deficient phase grows remains largely unknown. In this presentation, I will show our recent study on the structural dynamics in the ionic liquid-gated nickelate NdNiO3. Combining wide-angle X-ray photon correlation spectroscopy (XPCS) and X-ray reflection interface microscopy (XRIM), we were able to resolve the details of the oxygen migration process, such as dimensionality, reversibility, and morphology. I will also discuss the highly anisotropic oxygen migration process and how it relates to the strain state.

CONTROL ID: 3981513
CURRENT SYMPOSIUM: S12 - Magnetic and Superconducting Materials: From Basic Science to Applications
CURRENT SESSION: Other magnetic materials
TITLE: Combinatorial synthesis, crystal structure, and properties of magnetic and semiconducting $MnSnN_2$ thin films
PRESENTATION TYPE: Contributed (Oral)
PRESENTER (FIRST NAME ONLY): Christopher
PRESENTER (LAST NAME ONLY): Rom
PRESENTER (INSTITUTION ONLY): National Renewable Energy Laboratory
PRESENTER (EMAIL ONLY): christopher.rom@nrel.gov
PRESENTER (COUNTRY ONLY): United States
DATE/TIME SUBMITTED: September 14, 2023, 05:08 PM
ABSTRACT BODY:
Abstract Body: Magnetic semiconductors hold potential for spintronic applications, yet this class of materials remain underexplored. To discover new candidate materials, our team has used combinatorial sputtering to rapidly explore uncharted phase spaces. Here, we report on our recent discovery of $MnSnN_2$ via combinatorial sputtering. $MnSnN_2$ is a new semiconductor with antiferromagnetic correlations at low temperature. Our X-ray diffraction measurements characterize the material in detail across a wide composition, revealing cation disorder. Our magnetic measurements on $MnSnN_2$ reveal antiferromagnetic correlations below 10 K, which contrast dramatically with the high temperature antiferromagnetism of $MnSiN_2$ and $MnGeN_2$ that have been synthesized in bulk. We use Monte Carlo calculations to show how disorder lowers the magnetic transition temperature in $MnSnN_2$ compared to ordered analogs. These findings reveal how cation disorder can affect magnetism in these semiconducting materials, opening a pathway for precisely tuning the properties of new materials for microelectronics.

CONTROL ID: 3981830
CURRENT SYMPOSIUM: S18 - Structure and function in high-entropy, entropy-stabilized, and related oxides
CURRENT SESSION: Characterization and quantification of structure and properties
TITLE: Exploring new High-Entropy Oxides: Synthesizability and Mixability
PRESENTATION TYPE: Invited (by Invitation Only)

PRESENTER (FIRST NAME ONLY): Andrea
PRESENTER (LAST NAME ONLY): Kirsch
PRESENTER (INSTITUTION ONLY): University of Copenhagen
PRESENTER (EMAIL ONLY): anki@chem.ku.dk
PRESENTER (COUNTRY ONLY): Denmark

DATE/TIME SUBMITTED: September 18, 2023, 08:08 AM

ABSTRACT BODY:

Abstract Body: Since the introduction of the high-entropy concept in 2004, engineering compositional disorder into materials has gained significant attention. High-entropy oxides (HEOs) represent a new class of complex solid solutions featuring five or more metal cations in ca. equal amounts on a crystallographic site. Due to their compositional flexibility, they offer a wider design space than conventional solid solutions and hence more possibilities for property engineering. Remarkable properties have been reported for HEOs, among them ultralow thermal conductivity. HEOs have a strong tendency to crystallize in simple cubic crystal structures, and mainly rock salt, fluorite, and perovskite HEOs have been reported.
We focus on searching for new structure-types that can form as HEO and mapping out solution limits in the respective systems. As an example, we here report on the synthesis of new high-entropy mullite-type oxides. The material's crystal structure, microstructure and chemical homogeneity are investigated by a combination of structural, spectroscopic and microscopic techniques. Following the crystallization of the HEOs by coupled in situ XRD/XANES, we find that they form via an intermediate amorphous phase maintaining the homogeneity of the reaction mixture throughout the full synthesis process. Our synthesis has proven to be also efficient in the synthesis of many other HEOs, e.g. spinels and perovskites.

CONTROL ID: 3981684
CURRENT SYMPOSIUM: S7 - Defects and Transport in Ceramics
CURRENT SESSION: Defect mobility and transport behavior, including under operando or in extreme environments
TITLE: One stone, two birds: Using high electric fields to enhance the mobility and the concentration of point defects in ion-conducting solids
PRESENTATION TYPE: Contributed (Oral)
PRESENTER (FIRST NAME ONLY): Dennis
PRESENTER (LAST NAME ONLY): Kemp
PRESENTER (INSTITUTION ONLY): RWTH Aachen University
PRESENTER (EMAIL ONLY): kemp@pc.rwth-aachen.de
PRESENTER (COUNTRY ONLY): Germany
DATE/TIME SUBMITTED: September 18, 2023, 04:57 AM
ABSTRACT BODY:
Abstract Body: Improving the ionic conductivity of outstanding, composition-optimised crystalline electrolytes is a major challenge. Increases of orders of magnitude require, conceivably, highly non-linear effects. One known possibility is the use of high electric fields to increase the point-defect mobility.
In this study, we investigated a second possibility, that high electric fields can increase substantially point-defect concentrations. As a model system we took a pyrochlore oxide ($La_2Zr_2O_7$) for its combination of structural vacancies and dominant anti-Frenkel disorder; and we performed molecular-dynamics simulations with many-body potentials as a function of temperature and applied electric field. Results within the linear regime yielded the activation enthalpies and entropies of oxygen-vacancy and oxygen-interstitial migration, and from three independent methods, the enthalpy and entropy of anti-Frenkel disorder. Transport data for the non-linear regime were consistent with field-enhanced defect concentrations and defect mobilities. In particular, a route for separating the two effects was shown, and an analytical expression for the field-dependent anti-Frenkel equilibrium constant was derived.
In summary, we demonstrated that the one stone of a non-linear driving force can be used to hit two birds of defect behaviour.

CONTROL ID: 3981714

CURRENT SYMPOSIUM: S7 - Defects and Transport in Ceramics

CURRENT SESSION: Predictive bulk and interfacial point defect energetics and equilibria from density functional theory, molecular dynamics, and other computational methods

TITLE: A molecular dynamics study of oxygen diffusion in brownmillerite $Sr_2Fe_2O_5$

PRESENTATION TYPE: Contributed (Oral)

PRESENTER (FIRST NAME ONLY): Sonja

PRESENTER (LAST NAME ONLY): Ambaum

PRESENTER (INSTITUTION ONLY): RWTH Aachen University

PRESENTER (EMAIL ONLY): ambaum@pc.rwth-aachen.de

PRESENTER (COUNTRY ONLY): Germany

DATE/TIME SUBMITTED: September 19, 2023, 05:10 AM

ABSTRACT BODY:

Abstract Body: The topotactic phase transition of some perovskites, from perovskite $ABO_{3-\delta}$ to brownmillerite $A_2B_2O_5$, has recently attracted attention because of its potential for application in redox-induced resistive random access memories. The perovskite phase with its disordered vacancy distribution shows high oxide-ion mobility in contrast to the brownmillerite phase with its one-dimensionally ordered oxygen vacancies. The ion migration kinetics of the nanoscale redox processes are thus a key factor in determining the switching kinetics of topotactic memresistive devices. Oxygen transport in the brownmillerite phase is, however, poorly understood. Since applying voltage to thin films can result high electric fields it is, moreover, substantial to understand also the oxygen migration kinetics under the influence of an electric field.

In this study, we investigated oxygen transport in the brownmillerite $Sr_2Fe_2O_5$ system using molecular dynamics (MD) simulations with empirical pair potentials (EPP). Our simulations allowed us to study in detail the formation of oxygen vacancies and oxygen interstitials (anti-Frenkel disorder), from which we determined the enthalpy and entropy of anti-Frenkel disorder. MD simulations on $Sr_2Fe_2O_5$ with oxygen deficiency ($Sr_2Fe_2O_{5-x}$) or excess ($Sr_2Fe_2O_{5+x}$) yielded information on oxygen diffusion in terms of mechanisms and diffusion coefficients respectively oxygen ion mobilities.

CONTROL ID: 3981879

CURRENT SYMPOSIUM: S7 - Defects and Transport in Ceramics

CURRENT SESSION: High-throughput, screening, and combinatorial methods applied to the study of defect equilibria and point defect-mediated properties

TITLE: High-Throughput Optical Relaxation Analysis of Crystallization and Surface Exchange Kinetics on Combinatorial Mixed Conducting Films

PRESENTATION TYPE: Contributed (Oral)

PRESENTER (FIRST NAME ONLY): Alexia

PRESENTER (LAST NAME ONLY): Popescu

PRESENTER (INSTITUTION ONLY): University of Illinois at Urbana-Champaign

PRESENTER (EMAIL ONLY): alexia3@illinois.edu

PRESENTER (COUNTRY ONLY): United States

DATE/TIME SUBMITTED: September 25, 2023, 10:33 PM

ABSTRACT BODY:

Abstract Body: Crystallization and oxygen surface exchange kinetics are important parameters for the processing and use of mixed conducting perovskite oxides (ABO_3), but sequential synthesis and testing of compositions is time-consuming. By pulsed laser deposition of combinatorial $Sr(Ti_{0.65}Fe_{0.35-x}Co_x)O_{3-\delta}$ films with an Fe/Co composition gradient, and application of a novel 2D optical transmission relaxation (2D-OTR) method at various temperatures, we demonstrate high-throughput investigation of crystallization and exchange kinetics. Film libraries were grown through a multilayer approach from end-member targets, either amorphous at ~25 °C or crystalline at ~700 °C. The B-site lateral composition gradient and consistent thickness were confirmed by Rutherford backscattering spectrometry, while secondary ion mass spectrometry verified chemical homogeneity through the depth. Surface oxygen exchange coefficients (k_{chem}) and crystallization temperatures were determined with spatial resolution using 2D-OTR, which relates light transmission to defect concentration changes. Surface chemistry and morphology variations were analyzed across the libraries with X-ray photoelectron spectroscopy and atomic force microscopy, to understand the relationship between kinetics and deposition temperature and trends in k_{chem} and crystallization temperature vs. Fe/Co content.

CONTROL ID: 3982016
CURRENT SYMPOSIUM: S14 - Emerging Semiconductor Materials and Interfaces
CURRENT SESSION: 2D materials - synthesis, characterization, theory
TITLE: Two-Dimensional Neuromorphic Computing Materials and Devices
PRESENTATION TYPE: Invited (by Invitation Only)
PRESENTER (FIRST NAME ONLY): Mark
PRESENTER (LAST NAME ONLY): Hersam
PRESENTER (INSTITUTION ONLY): Northwestern University
PRESENTER (EMAIL ONLY): m-hersam@northwestern.edu
PRESENTER (COUNTRY ONLY): United States
DATE/TIME SUBMITTED: September 16, 2023, 06:31 PM
ABSTRACT BODY:
Abstract Body: The exponentially improving performance of digital computers has recently slowed due to the speed and power consumption issues resulting from the von Neumann bottleneck. In contrast, neuromorphic computing circumvents these limitations by spatially co-locating logic and memory in a manner analogous to biological neuronal networks. Beyond reducing power consumption, neuromorphic devices provide efficient architectures for image recognition, machine learning, and artificial intelligence. This talk will explore how 2D nanoelectronic materials enable gate-tunable neuromorphic devices. For example, by utilizing self-aligned, atomically thin heterojunctions, dual-gated Gaussian transistors have been realized, which show tunable anti-ambipolarity for artificial neurons, competitive learning, and mixed-kernel support vector machines. In addition, field-driven defect motion in polycrystalline monolayer MoS_2 enables gate-tunable memristive phenomena that serve as the basis of hybrid memristor/transistor devices (i.e., 'memtransistors') that concurrently provide logic and data storage functions. The planar geometry of memtransistors further allows multiple contacts and dual gating that mimic the behavior of biological systems such as heterosynaptic responses. Overall, this work introduces foundational circuit elements for neuromorphic computing by utilizing the unique quantum characteristics of 2D nanoelectronic materials.

CONTROL ID: 3982193
CURRENT SYMPOSIUM: S12 - Magnetic and Superconducting Materials: From Basic Science to Applications
CURRENT SESSION: Superconducting and related magnetic materials
TITLE: Machine learning and first-principles simulations of BCS superconductors
PRESENTATION TYPE: Invited (by Invitation Only)
PRESENTER (FIRST NAME ONLY): Cheng-Chien
PRESENTER (LAST NAME ONLY): Chen
PRESENTER (INSTITUTION ONLY): University of Alabama at Birmingham
PRESENTER (EMAIL ONLY): chencc@uab.edu
PRESENTER (COUNTRY ONLY): United States
DATE/TIME SUBMITTED: September 17, 2023, 06:12 PM
ABSTRACT BODY:
Abstract Body: A relationship between the Debye temperature Θ_D and the superconducting transition temperature T_c of conventional, phonon-mediated superconductors has been proposed recently [in npj Quantum Materials 3, 59 (2018)]. The relationship indicates that $T_c \leq A\Theta_D$ for BCS superconductors, with A of order ~0.1. Here, I will first discuss using machine learning models for predicting Θ_D to verify the proposal. It is found that the conventional superconductors in the NIMS SuperCon database indeed follow the proposed bound. The implication of our study for high-pressure hydrides will be discussed. First-principles calculations for robust superconducting phenomenon in alloying systems and potential higher-T_c conventional superconductors in superhard metals also will be presented.

CONTROL ID: 3982310
CURRENT SYMPOSIUM: S8 - Data-driven and model-supported structure-property relationships in complex electroceramics
CURRENT SESSION: Local structure and properties of complex electroceramics
TITLE: Functional Inorganics Driven by Immiscibility
PRESENTATION TYPE: Invited (by Invitation Only)
PRESENTER (FIRST NAME ONLY): David
PRESENTER (LAST NAME ONLY): Hall
PRESENTER (INSTITUTION ONLY): University of Manchester
PRESENTER (EMAIL ONLY): david.a.hall@manchester.ac.uk
PRESENTER (COUNTRY ONLY): United Kingdom
DATE/TIME SUBMITTED: September 18, 2023, 06:09 AM
ABSTRACT BODY:
Abstract Body: Perovskite-structured solid solutions based on bismuth ferrite show great potential for applications in dielectric, piezoelectric and electrocaloric devices. The processing procedures employed to synthesize such materials by solid state reaction exert a strong influence on their structure and properties, often resulting in core-shell type microstructural features. The factors controlling evolution of microstructure in $BiFeO_3$-$BaTiO_3$ (BF-BT) ceramics will be illustrated in terms of micro-chemical heterogeneity in calcined powders and sintered ceramics, using SEM, TEM and nano-SIMS, correlated with corresponding changes in the functional properties and quantitative analysis of ferroelectric domain switching behaviour.

BF-BT ceramics also exhibit remarkable improvements in ferroelectric ordering and functional properties in response to thermal treatment by quenching from high temperatures (>800 °C). Despite numerous investigations, the origin of these transformations has not yet been identified conclusively. In the present work, we focus on the effects of immiscibility between the BF and BT components. It is demonstrated that the development of phase separation during cooling plays a key role in the destabilisation of the ferroelectric rhombohedral phase and modification of dielectric/ferroelectric properties.

CONTROL ID: 3982378
CURRENT SYMPOSIUM: S15 - Electronics Materials for 5G/6G Applications
CURRENT SESSION: Materials growth and design
TITLE: Polymer Coated Ceramic Powders for Electronics Applications
PRESENTATION TYPE: Contributed (Oral)
PRESENTER (FIRST NAME ONLY): Christopher
PRESENTER (LAST NAME ONLY): Grabowski
PRESENTER (INSTITUTION ONLY): SABIC
PRESENTER (EMAIL ONLY): christopher.grabowski@sabic-hpp.com
PRESENTER (COUNTRY ONLY): United States
DATE/TIME SUBMITTED: September 22, 2023, 10:10 AM
ABSTRACT BODY:
Abstract Body: Composites made from polymer and ceramic materials serve as the dielectric component for many applications in the electronics industry. Balancing the dispersion, distribution and loading of ceramic fillers within a polymer matrix greatly affects key dielectric and thermal properties of the resulting composite structure as well as the ability to convert into parts using traditional plastic processing methods like compounding and injection molding. This talk reports on a class of 'moldable' powders that contain ceramic volume fractions difficult to achieve using traditional processing methods. We show that distributing ceramic content at loadings above 40% by volume enables a wider design space with higher dielectric permittivity, reduced CTE, and higher thermal conductivity. Ceramic powders, including TiO_2 and Al_2O_3, are shown to be compatible with a wide range of polymer resins to achieve tunable dielectric, thermal, and/or mechanical properties.

CONTROL ID: 3982401
CURRENT SYMPOSIUM: Plenary Session
CURRENT SESSION: None
TITLE: Towards sub-THz telecommunication: Advantages with ultra-low permittivity dielectrics
PRESENTATION TYPE: Plenary
PRESENTER (FIRST NAME ONLY): Heli
PRESENTER (LAST NAME ONLY): Jantunen
PRESENTER (INSTITUTION ONLY): University of Oulu
PRESENTER (EMAIL ONLY): heli.jantunen@oulu.fi
PRESENTER (COUNTRY ONLY): Finland
DATE/TIME SUBMITTED: September 18, 2023, 10:07 AM
ABSTRACT BODY:
Abstract Body: Ultra-low permittivity materials are of paramount importance for 6G telecommunication systems. When the low losses are reached at the same time, they enable the efficient propagation of signals at extremely high frequencies, reduce signal loss, enhance antenna performance, and contribute to overall efficiency and reliability. As the demand for faster and more reliable wireless communication continues to grow, the development and utilization of these materials will be critical to realizing the full potential of 6G technology.
This lecture will review the advantages the low-loss and low-permittivity dielectrics can provide as well as the current situation with the materials available. Some recent approaches for low-loss and low-permittivity materials for sub-THz frequencies to be presented involve ceramics, polymers, and wood-based composites for antenna substrates and radio lenses with permittivity values well under 4 and losses ~10^{-3}. The importance of surface smoothening to decrease the air-lens reflection to improve the antenna gain, the achievable total transmittance, and the fabrication methods to realize different lens shapes are also discussed.

CONTROL ID: 3982439
CURRENT SYMPOSIUM: S2 - Advanced Microelectronics
CURRENT SESSION: Dielectrics and ferroelectrics for device applications
TITLE: Ferroelectric switching kinetics in epitaxial ferroelectric thin films
PRESENTATION TYPE: Invited (by Invitation Only)
PRESENTER (FIRST NAME ONLY): John
PRESENTER (LAST NAME ONLY): Heron
PRESENTER (INSTITUTION ONLY): University of Michigan
PRESENTER (EMAIL ONLY): jtheron@umich.edu
PRESENTER (COUNTRY ONLY): United States
DATE/TIME SUBMITTED: September 18, 2023, 10:49 AM
ABSTRACT BODY:
Abstract Body: Ferroelectrics are emerging materials that promise energy-efficient memory and compute devices. To impact microelectronics, however, the speed of ferroelectric operation at device scale (lateral: 10-100s nm, thickness: 10s nm) must be fast, 1-2 ns or preferably better. Here we examine ferroelectric switching in epitaxial ferroelectric thin film capacitors with 5 – 80 nm thickness and 300 nm – 40 microns lateral dimension. For PMN-PT films, we find that the ferroelectric switching behavior obeys the Kolmogorov–Avrami–Ishibashi theory. The switching kinetics is dominated by a mix of 2D and 3D domain growth. As a result, the ferroelectric switching time versus lateral scaling follows Mertz law down to at least 800 nm diameter capacitors indicating domain growth is the dominant switching mechanism in this range. Switching times for the smallest capacitors are in the few 100 ps range. Piezoresponse force microscopy measurements of 150 nm thick films reveal a clamped piezoresponse for capacitors above 800 nm with relaxed clamping near the perimeter. Capacitors below 400 nm in diameter are relaxed. Our results give a promising outlook for PMN-PT based ferroelectric and multiferroic devices for energy efficient logic and memory.

CONTROL ID: 3982456
CURRENT SYMPOSIUM: S15 - Electronics Materials for 5G/6G Applications
CURRENT SESSION: Materials growth and design
TITLE: Emerging Materials and Design Principles for Wurtzite-type Ferroelectrics
PRESENTATION TYPE: Contributed (Oral)
PRESENTER (FIRST NAME ONLY): Cheng-Wei
PRESENTER (LAST NAME ONLY): Lee
PRESENTER (INSTITUTION ONLY): Colorado School of Mines
PRESENTER (EMAIL ONLY): clee2@mines.edu
PRESENTER (COUNTRY ONLY): United States
DATE/TIME SUBMITTED: September 25, 2023, 11:31 PM
ABSTRACT BODY:
Abstract Body: Historically, octahedrally-bonded oxide perovskites have been the focus of ferroelectrics (FE) research. However, oxide thin films are plagued by defects, partly due to their structural incompatibility with microelectronic semiconductors such as Si, GaN etc., that are tetrahedrally bonded. Tetrahedral FEs are interesting because of their potential for direct integration with conventional microelectronics and unusual switching behavior; such interest has taken off since publication of the first report of ferroelectric (Al,Sc)N allloys in 2019. Subsequently, FE switching was demonstrated in (Al,B)N, (Ga,Sc)N, and (Zn,Mg)O alloys. Beyond the existence of switching and some composition trends, little is known about tetrahedral FEs. One major challenge for currently known tetrahedral FEs is that the coercive field required to switch the polarization is several MV/cm, which is dangerously close to the breakdown fields. Instead of further engineering these alloys, we computationally explore multinary wurtzite-type compounds for promising candidates with low switching barriers and high dielectric breakdown fields. We identify several promising candidates for future experimental verification. We derive design principles based on metal-anion bond ionicity and lattice softness to lower switching barriers. Finally, we offer new insights into the polarization switching mechanism in multinary wurtzite-type compounds.

CONTROL ID: 3982491
CURRENT SYMPOSIUM: S5 - Controlling Grain Boundary Structure, Chemistry, and their Network as a Function of Mtl Processing
CURRENT SESSION: Microstructure evolution
TITLE: Electric Current Effects on the Plastic Deformation Behavior and Grain Morphology of Pure Copper, Iron, and Titanium
PRESENTATION TYPE: Invited (by Invitation Only)
PRESENTER (FIRST NAME ONLY): Christopher
PRESENTER (LAST NAME ONLY): Rudolf
PRESENTER (INSTITUTION ONLY): US Naval Research Laboratory
PRESENTER (EMAIL ONLY): christopher.rudolf@nrl.navy.mil
PRESENTER (COUNTRY ONLY): United States
DATE/TIME SUBMITTED: September 21, 2023, 11:37 AM
ABSTRACT BODY:
Abstract Body: Uniaxial tension tests were performed on pure polycrystalline copper, iron, and titanium specimens with various applied constant (dc) current levels and at zero current with temperature histories matched to the current tests. The experiments achieved uniform strain, current density, and temperature conditions along the specimen gage length for unambiguous interpretation of the test data. The results showed non-thermal current effects only with the titanium; 20% reduction in ultimate strength relative to the temperature-matched strength was observed as well as significant inhomogeneous grain growth. No discernable changes in microstructure were observed in specimens deformed at matching temperatures or with applied current but no deformation (matching temperatures). The electron-wind and local Joule heating mechanisms for EAD do not produce effects large enough to explain the observed titanium results. Dislocation scattering by thermal phonons and electrons associated with the radial and axial heat fluxes generated in the titanium tensile specimens with bulk Joule heating is proposed as a new candidate EAD mechanism. The experimental results and possible link to thermal phonon/electron scattering suggests several new avenues of research on EAD of metals.

CONTROL ID: 3982554

CURRENT SYMPOSIUM: S4 - Complex Oxide Thin Films and Heterostructures: From Synthesis to Strain/Interface-engineered Emergent Properties

CURRENT SESSION: Controlled synthesis of epitaxial thin films and nanocomposites, including heterostructures, superlattices, and vertically aligned nanocomposites for functional properties

TITLE: Composition-driven structural and dielectric transitions in epitaxial (1-x)$PbMg_{1/3}Nb_{2/3}O_3$-(x)$PbZrO_3$ thin films

PRESENTATION TYPE: Contributed (Oral)

PRESENTER (FIRST NAME ONLY): Sreekeerthi

PRESENTER (LAST NAME ONLY): Pamula

PRESENTER (INSTITUTION ONLY): University of California Berkeley

PRESENTER (EMAIL ONLY): sreekeerthi67@berkeley.edu

PRESENTER (COUNTRY ONLY): United States

DATE/TIME SUBMITTED: September 25, 2023, 04:07 PM

ABSTRACT BODY:

Abstract Body: Antiferroelectrics show a reversible phase transition to ferroelectric order and a double polarization–electric field hysteresis loop and are promising for dielectric and electromechanical applications. Research has focused on alloying antiferroelectrics with ferroelectrics and paraelectrics, but less so with relaxor ferroelectrics. Such solid solutions are interesting due to the unique properties of relaxor ferroelectrics including their robust saturation polarization, high dielectric tunability, and low hysteresis loss. Here, solid solutions of (1-x)$PbMg_{1/3}Nb_{2/3}O_3$-(x)$PbZrO_3$ in the range $0.85 \leq x \leq 1$ were grown as epitaxial thin films on $SrTiO_3$ (001) substrates. An increasing dielectric response and tunability was observed as the system transitions from antiferroelectric to a ferroelectric phase with a phase boundary near x = 0.92. Among other interesting observations, these films had twice the breakdown field of equally thick $PbZrO_3$ films, leakage current two orders of magnitude lower, with similar dielectric and electromechanical properties. Leakage reduction is attributed to the associated defect formation arising from alloying. This enhances the electromechanical response and energy-storage efficiency as large fields can be applied. On the whole, this study provides insights into the behavior of this antiferroelectric-relaxor system and the underlying mechanisms.

CONTROL ID: 3982586
CURRENT SYMPOSIUM: S2 - Advanced Microelectronics
CURRENT SESSION: Emergent devices for novel computing such as neuromorphic computing
TITLE: Neuromorphic Computing: Bridging the gap between Nanoelectronics, Neuroscience and Machine Learning
PRESENTATION TYPE: Invited (by Invitation Only)
PRESENTER (FIRST NAME ONLY): Abhronil
PRESENTER (LAST NAME ONLY): Sengupta
PRESENTER (INSTITUTION ONLY): Pennsylvania State University
PRESENTER (EMAIL ONLY): sengupta@psu.edu
PRESENTER (COUNTRY ONLY): United States
DATE/TIME SUBMITTED: September 22, 2023, 07:08 PM
ABSTRACT BODY:
Abstract Body: While research in designing brain-inspired algorithms have attained a stage where such Artificial Intelligence platforms are being able to outperform humans at several cognitive tasks, an often-unnoticed cost is the huge computational expenses required for running these algorithms in hardware. Recent explorations have also revealed several algorithmic vulnerabilities of deep learning systems like adversarial susceptibility, lack of explainability, catastrophic forgetting, to name a few. Bridging the computational and algorithmic efficiency gap necessitates the exploration of hardware and algorithms that provide a better match to the computational primitives of biological processing – neurons and synapses, and which require a significant rethinking of traditional von-Neumann based computing. This talk reviews recent developments in the domain of neuromorphic computing paradigms from an overarching system science perspective with an end-to-end co-design focus from computational neuroscience and machine learning to hardware and applications. Such neuromorphic systems can potentially provide significantly lower computational overhead in contrast to standard deep learning platforms, especially in sparse, event-driven application domains with temporal information processing.

CONTROL ID: 3982639
CURRENT SYMPOSIUM: S15 - Electronics Materials for 5G/6G Applications
CURRENT SESSION: Materials growth and design
TITLE: Enabling Thin Glass for High Frequency Electronic Applications
PRESENTATION TYPE: Invited (by Invitation Only)
PRESENTER (FIRST NAME ONLY): David
PRESENTER (LAST NAME ONLY): Levy
PRESENTER (INSTITUTION ONLY): Mosaic Microsystems
PRESENTER (EMAIL ONLY): david.levy@mosaicmicro.com
PRESENTER (COUNTRY ONLY): United States
DATE/TIME SUBMITTED: September 25, 2023, 06:59 PM
ABSTRACT BODY:
Abstract Body: Glass has excellent electrical and mechanical properties as a substate for advanced electronic packaging, and even photonics applications. It has a large bulk resistance and low loss at high frequency. This is particularly true of high purity fused silica (FS), which exhibits a very low loss tangent at GHz frequencies. At the same time, glass is moisture insensitive and dimensionally stable, properties that allow it to be inserted in many of the processing technologies reserved for silicon.

We discuss a novel through glass via (TGV) process, and the formation of full interposers on thin glass. We demonstrate capability to produce void-free, hermetic, copper-filled vias, ranging in aspect ratio from 2:1 to 8:1, in addition to redistribution layer (RDL) fabrication with up to three metal redistribution layers per side.

Our technology has also enabled the fabrication of very thin glass with tight total thickness variation (TTV), which allows good uniformity of devices across the whole substrate. In addition to free-standing glass with thickness of less than 100 microns, we have demonstrated supported fused silica substrates with a thickness as low as 7 um, fabricated from bulk fused silica starting material.

CONTROL ID: 3982642
CURRENT SYMPOSIUM: S18 - Structure and function in high-entropy, entropy-stabilized, and related oxides
CURRENT SESSION: Understanding and predicting structure and properties through computation and data science
TITLE: Understanding the limits to Short-range order Suppression in Many-Component Disordered Rock Salt Lithium-ion Cathode Materials
PRESENTATION TYPE: Invited (by Invitation Only)
PRESENTER (FIRST NAME ONLY): Alex
PRESENTER (LAST NAME ONLY): Squires
PRESENTER (INSTITUTION ONLY): University of Birmingham
PRESENTER (EMAIL ONLY): a.squires@bham.ac.uk
PRESENTER (COUNTRY ONLY): United Kingdom
DATE/TIME SUBMITTED: September 18, 2023, 04:42 PM
ABSTRACT BODY:
Abstract Body: Disordered rock salt cathode materials are perceived as excellent candidates for application as future lithium-ion cathode materials enabling novel sustainable battery chemistries. It is found that lithium diffusion in these materials is suppressed by their short range order (SRO) preferences. Suppressing this SRO is therefore crucial to maximising their performance. Alloying many transition metals together across the cation sublattice, creating so-called high-entropy rock salts, has been proposed as a method for minimising SRO. This poses an exciting materials design challenge which requires developing an understanding of the effects of cation sublattice configurational entropy and how this influences the SRO in as-synthesised cathode materials. In this work we have developed a cluster-expansion model to drive Monte Carlo simulations examining the connectivity of the Li percolation network when varying the composition of high-component DRX Li cathode materials. We reveal the relative importance of the specific composition of the system as compared to a general increase in the configurational entropy and generate and explain compositional maps for developing the optimal SRO suppression in many-component disordered rock salt oxyfluoride cathode materials, guiding experimental collaborators to optimal compositions.

CONTROL ID: 3982718
CURRENT SYMPOSIUM: S2 - Advanced Microelectronics
CURRENT SESSION: Emergent devices for novel computing such as neuromorphic computing
TITLE: Variation-Aware FeFET-Based In-Memory Computing Leveraging Probabilistic Deep Learning
PRESENTATION TYPE: Contributed (Oral)
PRESENTER (FIRST NAME ONLY): Bibhas
PRESENTER (LAST NAME ONLY): Manna
PRESENTER (INSTITUTION ONLY): The Pennsylvania State University
PRESENTER (EMAIL ONLY): bxm5752@psu.edu
PRESENTER (COUNTRY ONLY): United States
DATE/TIME SUBMITTED: September 22, 2023, 08:35 PM
ABSTRACT BODY:
Abstract Body: Reliability issues stemming from device level non-idealities of ferroelectric field-effect transistors (FeFET), especially at scaled dimensions, cause substantial degradation in the accuracy of In-Memory crossbar based AI systems. In this work, we present a variation-aware design technique to characterize the device level variations and to mitigate their impact on hardware accuracy employing a Bayesian neural network (BNN) approach. An effective conductance variation model was derived from the experimental measurements of cycle-to-cycle (C2C) and device-to-device (D2D) variations performed on FeFET devices fabricated using 28 nm high-k metal gate technology. The variations were found to be a function of different conductance states within the given programming range, which sharply contrasts earlier efforts where a fixed variation model was considered for all conductance values. Such variation characteristics formulated for three device sizes at different read voltages were provided as prior variation information to the BNN to yield a more exact and reliable inference. Near-ideal accuracy (~82%) of AlexNet/CIFAR10, almost insensitive to device sizes and read voltages, demonstrates the efficacy of our proposed device-algorithm co-design technique.

CONTROL ID: 3982858

CURRENT SYMPOSIUM: S4 - Complex Oxide Thin Films and Heterostructures: From Synthesis to Strain/Interface-engineered Emergent Properties

CURRENT SESSION: Controlled synthesis of epitaxial thin films and nanocomposites, including heterostructures, superlattices, and vertically aligned nanocomposites for functional properties

TITLE: Growth of crystalline $SrVO_3$ thin films on glass: Enabling applications as Transparent Conducting Oxide

PRESENTATION TYPE: Contributed (Oral)

PRESENTER (FIRST NAME ONLY): Ulrike

PRESENTER (LAST NAME ONLY): Lüders

PRESENTER (INSTITUTION ONLY): CRISMAT

PRESENTER (EMAIL ONLY): ulrike.luders@ensicaen.fr

PRESENTER (COUNTRY ONLY): France

DATE/TIME SUBMITTED: September 25, 2023, 05:06 AM

ABSTRACT BODY:

Abstract Body: $SrVO_3$ (SVO) was identified as a new, indium-free, and highly performing transparent conducting oxide, comparable to indium-tin-oxide (ITO). However, these performances are only reached when the SVO thin films have the perovskite structure, amorphous films being insulating. For applications in devices, the deposition of conducting, and therefore crystalline, SVO films on glass is necessary, but notoriously difficult. We have explored two approaches to achieve crystallized SVO films on glass by using binary oxide buffer layers or oxide 2D nanosheets.

Film deposition was carried out by Pulsed Laser Deposition. In the case of the binary oxide buffer layers, the SVO films are polycrystalline, reaching resistivity values of 4×10^{-4} Ωcm depending on the nature of the glass substrate and the used buffer layer. These values, about a factor of 4 higher than observed for monocrystalline SVO films, are still very promising for applications, especially in relation to the optical transparency above 70 %. Second, using $[Ca_2Nb_3O_{10}]^-$ nanosheets, (001) textured SVO films are obtained. Showing only a small influence of the grain boundaries, these films allow to reach resistivity values in the same range. Interestingly, compared to polycrystalline films, the thickness of the films on nanosheets can be strongly reduced, inducing a higher transparency without compromising the transport properties.

CONTROL ID: 3983129
CURRENT SYMPOSIUM: S2 - Advanced Microelectronics
CURRENT SESSION: High performance thin film and 2D transistors/devices
TITLE: Measuring the phonon contributions to total thermal conductivity of Ru and W thin films using a SSTR technique
PRESENTATION TYPE: Contributed (Oral)
PRESENTER (FIRST NAME ONLY): Md Rafiqul
PRESENTER (LAST NAME ONLY): Islam
PRESENTER (INSTITUTION ONLY): University of Virginia
PRESENTER (EMAIL ONLY): mi9jm@virginia.edu
PRESENTER (COUNTRY ONLY): United States
DATE/TIME SUBMITTED: September 19, 2023, 05:17 PM
ABSTRACT BODY:
Abstract Body: The total thermal conductivity of a metal is the sum of electronic thermal conductivity and phonon thermal conductivity, along with any other heat carriers if they exist. Typically we assume that phonon thermal conductivity is negligible for metals, however, this hypothesis has not been subjected to stringent experimental validation due to difficulties in measuring phonon thermal conductivity. In this work, we study the phonon and electron contribution to the thermal conductivity of ruthenium (Ru) and tungsten (W) films via independent measurements of the thermal and electrical transport properties. We perform sheet resistance and thermoreflectance-based nanoscale thermal conductance measurements on Ru films ranging from 20 to 100 nm thick and W films ranging from 20 to 30 nm thick. We obtain the electrical resistivity with the four-point probe method and then calculate in-plane thermal conductivity employing the Wiedemann-Franz (WF) law. We directly measure the in-plane thermal conductivity of the Ru and W thin films via a steady-state thermoreflectance technique. The in-plane thermal conductivity of these thin films measured with thermoreflectance-based methods is at least 15% higher than the WF law-derived one. This indicates the viability of thermoreflectance-based method in capturing the phonon contributions to the total thermal conductivity of metals.

CONTROL ID: 3983152
CURRENT SYMPOSIUM: S10 - Fundamental mechanisms in materials for next generation computing devices
CURRENT SESSION: Advanced characterization techniques
TITLE: Probing nanoscale electrical memory effects in multi-well ferroelectrics and 2D semiconductors
PRESENTATION TYPE: Invited (by Invitation Only)
PRESENTER (FIRST NAME ONLY): Sabine
PRESENTER (LAST NAME ONLY): Neumayer
PRESENTER (INSTITUTION ONLY): Oak Ridge National Lab
PRESENTER (EMAIL ONLY): neumayersm@ornl.gov
PRESENTER (COUNTRY ONLY): United States
DATE/TIME SUBMITTED: September 25, 2023, 04:07 PM
ABSTRACT BODY:
Abstract Body: The increasing importance of artificial intelligence and big data applications demands the development of energy efficient computing paradigms. The high power demand of conventional computing arises from the need to transfer data between different locations for processing and storage, resulting in redundant energy consumption. Neuromorphic technologies aim to overcome these limitations by co-locating storage and computing in the same physical space. In particular, the possibility to encode data directly into material properties is desired for neuromorphic hardware. Electrical scanning probe microscopy (SPM) is an effective tool to characterize the functionality of materials on the micro-and nanoscale and evaluate them regarding their neuromorphic capabilities. In this talk, I will showcase nanoscale impedance switching and electrical memory effects based on different mechanisms in multi-well ferroelectrics and 2D layered semiconductors. Multimodal SPM revealed mem-capacitive effects in $Sn_2P_2S_6$ that arise from metastable polarization states. In semiconducting $PdSe_2$, Se vacancies lead to nanoscale capacitive and resistive switching, which was detected using scanning microwave impedance microscopy. Support by CNMS which is a US DOE User Facility at ORNL. Part of this research was sponsored by the LDRD program of ORNL, managed by UT-Battelle, LLC, for the US DOE.

CONTROL ID: 3983157

CURRENT SYMPOSIUM: S4 - Complex Oxide Thin Films and Heterostructures: From Synthesis to Strain/Interface-engineered Emergent Properties

CURRENT SESSION: Controlled synthesis of epitaxial thin films and nanocomposites, including heterostructures, superlattices, and vertically aligned nanocomposites for functional properties

TITLE: Enhanced Performance of CeO_2-Based Memristors Through Integration with Co-CeO2 Vertically Aligned Nanocomposite Thin Film for Neuromorphic Computing

PRESENTATION TYPE: Contributed (Oral)

PRESENTER (FIRST NAME ONLY): Zedong

PRESENTER (LAST NAME ONLY): Hu

PRESENTER (INSTITUTION ONLY): Purdue University

PRESENTER (EMAIL ONLY): hu826@purdue.edu

PRESENTER (COUNTRY ONLY): United States

DATE/TIME SUBMITTED: September 25, 2023, 03:06 PM

ABSTRACT BODY:

Abstract Body: Cerium oxide (CeO2) based memristor, as one of the most promising non-volatile memory to address the ever-expanding computational demands in the realm of artificial intelligence (AI) and machine learning (ML), has garnered substantial attention owing to its fast switching speed, high-order dynamics, high stability, and long-term retention. While these inherent merits render them appealing for applications in neuromorphic computing, the integration of CeO2-based memristors has been hindered by inherent limitations, including instability, high switching voltages, and a low on/off ratio. In this study, we combined the conventional CeO2-based memristor with a novel Co-CeO2 vertically aligned nanocomposite (VAN) thin film and proposed a new device architecture denoted as CeO2-Co / CeO2 / Nb: STO. Co nanopillars provide more interfaces favoring the formation of oxygen vacancies, and function as field extractors. In summary, the on/off ratio has been increased by more than tenfold, the switching voltage has been reduced by over 30%, and the overall stability of the device has witnessed a substantial improvement, which makes it a highly promising candidate for future applications in neuromorphic computing.

CONTROL ID: 3983206

CURRENT SYMPOSIUM: S15 - Electronics Materials for 5G/6G Applications

CURRENT SESSION: 5G/6G Metrology

TITLE: Measuring out-of-plane permittivity of dielectric thin films in the millimeter-wave range

PRESENTATION TYPE: Contributed (Oral)

PRESENTER (FIRST NAME ONLY): Samuel

PRESENTER (LAST NAME ONLY): Freed

PRESENTER (INSTITUTION ONLY): University of Maryland

PRESENTER (EMAIL ONLY): sfreed@umd.edu

PRESENTER (COUNTRY ONLY): United States

DATE/TIME SUBMITTED: September 21, 2023, 03:32 PM

ABSTRACT BODY:

Abstract Body: Accurate permittivity data is critical for the design of next-generation microelectronics. Whether for quantum computing or advanced communications, industry needs to measure broadband, out-of-plane permittivity up to millimeter-wave frequencies. The current state-of-the-art for measuring out-of-plane permittivity of thin films involves either scanning microwave microscopy, which has limited accuracy, or a metal-insulator-metal (MIM) capacitor method, which is limited by large uncertainty at frequencies above 5 GHz. Here, I will discuss our approach to extend the MIM capacitor method to millimeter-wave frequencies. First, I will discuss how we built on the existing method to remove parasitics, increase accuracy, and extend the applicable frequency range by updating the device design. Next, I will provide insight into how we used both the already well-known silicon nitride and newly developed barium strontium titanate Ruddlesden-Popper thin films to validate the modified method. Finally, I will explore the computer modeling applications we developed to assist in the classification and analysis of new thin-film materials. These materials provide a case study for extracting out-of-plane permittivity at millimeter-wave frequencies.

CONTROL ID: 3983254
CURRENT SYMPOSIUM: S12 - Magnetic and Superconducting Materials: From Basic Science to Applications
CURRENT SESSION: Other magnetic materials
TITLE: Pressure-Tuned Structural, Electronic, Magnetic, and Optical Properties of MPX_3 (X = S, Se)
PRESENTATION TYPE: Invited (by Invitation Only)
PRESENTER (FIRST NAME ONLY): Takahiro
PRESENTER (LAST NAME ONLY): Matsuoka
PRESENTER (INSTITUTION ONLY): University of the Philippines Diliman
PRESENTER (EMAIL ONLY): tmatsuoka@up.edu.ph
PRESENTER (COUNTRY ONLY): Philippines
DATE/TIME SUBMITTED: September 21, 2023, 09:26 PM
ABSTRACT BODY:
Abstract Body: The MPX_3 (M = V, Mn, Fe, Co, Ni, Cd, and X = S, Se) are a family of two-dimensional (2D) antiferromagnetic van der Waals compounds. These compounds have been extensively studied in recent years due to their unique electronic structure and intriguing physical and chemical properties. Pressure is a highly effective external stimulus for tuning the properties and driving new states of matter in MPX_3. Under compression, these compounds exhibit interlayer sliding, insulator-to-metal transitions with significant volume reduction and spin state collapse, piezochromism, potential polar metal, orbital Mott phases, and superconductivity. This talk will discuss recent experimental and theoretical studies on MPX_3 under compression, with a focus on the structural, magnetic, and optical properties. It will also provide an outlook for future research on this family of compounds using strains.

CONTROL ID: 3983284
CURRENT SYMPOSIUM: S16 - New Frontiers and Science in Additive Manufacturing of Ceramic Materials
CURRENT SESSION: New approaches for AM processes of ceramics and ceramic-matrix composites
TITLE: Direct selective laser sintering of high-entropy carbide ceramics
PRESENTATION TYPE: Contributed (Oral)
PRESENTER (FIRST NAME ONLY): Bai
PRESENTER (LAST NAME ONLY): Cui
PRESENTER (INSTITUTION ONLY): University of Nebraska–Lincoln
PRESENTER (EMAIL ONLY): bcui3@unl.edu
PRESENTER (COUNTRY ONLY): United States
DATE/TIME SUBMITTED: September 26, 2023, 01:09 AM
ABSTRACT BODY:
Abstract Body: Direct selective laser sintering (SLS) process was successfully demonstrated for additive manufacturing of high-entropy carbide ceramics (HECC), in which a Yb fiber laser was employed for ultrafast (in seconds) reactive sintering of HECC specimens from a powder mixture of constitute monocarbides. A single-phase non-equiatomic HECC was successfully formed in the 4-HECC specimen with a uniform distribution of Zr, Nb, Hf, Ta, and C. In contrast, a three-layer microstructure was formed in the 5-HECC specimen with five metal elements (Zr, Nb, Hf, Ta and Ti), consisting of a TiC-rich top layer, a Zr-Hf-C enriched intermediate layer, and a non-equiatomic Zr-Ta-Nb-Hf-C HECC layer. Vickers hardness of 4- and 5-HECC specimens were 22.2 and 21.8 GPa, respectively, on the surface. These findings have important implications on the fundamental mechanisms governing interactions between continuous-wave lasers and monocarbide powders to form a solid solution of HECCs during SLS.

CONTROL ID: 3983288

CURRENT SYMPOSIUM: S2 - Advanced Microelectronics

CURRENT SESSION: Theory, modeling and first principles calculations of devices performance and properties

TITLE: Comparative Analysis of Gate-All-Around Nanowire Field Effect Transistor for Future Sensors Using Non-Equilibrium Green's Function

PRESENTATION TYPE: Contributed (Oral)

PRESENTER (FIRST NAME ONLY): Suruchi

PRESENTER (LAST NAME ONLY): Saini

PRESENTER (INSTITUTION ONLY): Kurukshetra University

PRESENTER (EMAIL ONLY): suruchi.electsc@kuk.ac.in

PRESENTER (COUNTRY ONLY): India

DATE/TIME SUBMITTED: September 23, 2023, 06:34 AM

ABSTRACT BODY:

Abstract Body: In line with the fast progress of nanotechnology, it is important to have compact semiconductor devices to function and offer the best results at various technology nodes. Gate-all-around (GAA) nanowire Field Effect Transistor (FET) is one of the most promising structures. A nanowire having an 8 nm channel length with the same source and drain doping concentration is used for this analysis. This work uses $sp^3d^5s^*$ tight-binding model for comparative analysis of Nanowire FET. The non-equilibrium Green's function (NEGF) approach is used for this simulation. The comparative analysis is made by changing the diameter of the nanowire, crystal orientation, and oxide thickness for different channel materials Si and GaAs. The results show that the GAA Si Nanowire is superior with a high current on-off ratio of the order of 10^{11} having a diameter of 1.5 nm. Also, the nanowire with a small diameter shows the best result in terms of On-current and Off-current. For crystal orientation x[100] which is along the transport of carriers in nanowire, the results obtained are very good as compared to other orientations. The hydrostatic strain has been analyzed. The change in the current on-off ratio with hydrostatic strain makes them a leading option for future sensors.

CONTROL ID: 3983376
CURRENT SYMPOSIUM: S4 - Complex Oxide Thin Films and Heterostructures: From Synthesis to Strain/Interface-engineered Emergent Properties
CURRENT SESSION: Controlled synthesis of epitaxial thin films and nanocomposites, including heterostructures, superlattices, and vertically aligned nanocomposites for functional properties
TITLE: Formation and transport of interfacial charges for neuromorphic microchips
PRESENTATION TYPE: Invited (by Invitation Only)
PRESENTER (FIRST NAME ONLY): Ho Nyung
PRESENTER (LAST NAME ONLY): Lee
PRESENTER (INSTITUTION ONLY): Oak Ridge National Lab
PRESENTER (EMAIL ONLY): hnlee@ornl.gov
PRESENTER (COUNTRY ONLY): United States
DATE/TIME SUBMITTED: September 20, 2023, 09:16 AM
ABSTRACT BODY:
Abstract Body: Complex oxides are known to possess the full spectrum of fascinating properties, including magnetism, colossal magnetoresistance, superconductivity, ferroelectricity, ionic conductivity, and more. The breadth of remarkable properties is the consequence of strong coupling among charge, spin, orbital, and lattice degrees of freedom. Spurred by recent advances in the synthesis of such artificial materials at the atomic scale, the physics of oxide heterostructures containing atomically smooth layers of such correlated electron materials with abrupt interfaces is a rapidly growing area. In this talk, examples of ionic interfaces studied by scanning transmission electron microscopy/electron energy-loss spectroscopy and atom probe topography will be presented, highlighting the importance of precision synthesis for heterostructuring, interfacing, and straining. The main topics include (1) strain control of orbital polarization, oxygen mobility, and cationic distribution in perovskite oxides and (2) microscopic understanding of interfacial charge and ion transport in a fluorite-bixbyite heterostructure, in which the well-defined interface provides an excellent pathway for ion transport critical for developing neuromorphic circuits and ionotronics.

CONTROL ID: 3983431
CURRENT SYMPOSIUM: S5 - Controlling Grain Boundary Structure, Chemistry, and their Network as a Function of Mtl Processing
CURRENT SESSION: Defect and segregation behavior
TITLE: A molecular-dynamics study of oxygen diffusion in polycrystalline $(La,Sr)FeO_3$
PRESENTATION TYPE: Contributed (Oral)
PRESENTER (FIRST NAME ONLY): Alexander
PRESENTER (LAST NAME ONLY): Bonkowski
PRESENTER (INSTITUTION ONLY): RWTH Aachen University
PRESENTER (EMAIL ONLY): alexander.bonkowski@rwth-aachen.de
PRESENTER (COUNTRY ONLY): Germany
DATE/TIME SUBMITTED: September 22, 2023, 08:14 AM
ABSTRACT BODY:
Abstract Body: Faster grain-boundary diffusion of oxygen has been observed experimentally in polycrystalline samples of Fe-based perovskite oxides at low temperatures, but this behaviour is at present not well understood. In this study we employed molecular-dynamics (MD) simulations with empirical force fields [1] to investigate oxygen transport in polycrystalline $La_{1-x}Sr_xFeO_{3-x/2}$ (x = 0.1 and 0.4) at temperatures $1000 \leq T / K \leq 2000$. Oxygen tracer diffusion coefficients obtained for both orthorhombic $La_{0.9}Sr_{0.1}FeO_{2.95}$ and cubic $La_{0.6}Sr_{0.4}FeO_{2.8}$ are compared with experimental data, and are found to give a satisfactory description of the oxygen transport behaviour as a function of T and x. Subsequently, we examined oxygen tracer diffusion in polycrystalline cells of $La_{1-x}Sr_xFeO_{3-x/2}$ with different grain boundary concentrations. In particular, the effects of different oxygen nonstoichiometries δ and of equilibrium (as opposed to random) defect distributions were examined. Our results indicate that the disrupted structures of the grain boundaries hinder oxygen diffusion; that Sr accumulation within grain-boundary regions does not produce faster diffusion; but that faster grain-boundary diffusion is observed when δ is decreased substantially.
[1] M. Cherry, et al., J. Solid State Chem. 1995, 118, 125–132; M. S. Islam, et al., J. Solid State Chem. 1996, 124, 230–237.

CONTROL ID: 3983472
CURRENT SYMPOSIUM: S11 - In situ/operando Characterization of Nanomaterials
CURRENT SESSION: Study of electronic materials such as Mott insulators, dielectrics, ionics, ferroics, semiconductors, superconductors etc. using advanced characterization techniques
TITLE: What lies beyond: Exploring ferroelectric domains beneath the surface with Tomographic AFM
PRESENTATION TYPE: Contributed (Oral)
PRESENTER (FIRST NAME ONLY): Karla
PRESENTER (LAST NAME ONLY): Del Cid-Ledezma
PRESENTER (INSTITUTION ONLY): University of Connecticut
PRESENTER (EMAIL ONLY): karla.del_cid-ledezma@uconn.edu
PRESENTER (COUNTRY ONLY): United States
DATE/TIME SUBMITTED: September 20, 2023, 12:29 PM
ABSTRACT BODY:
Abstract Body: This research presents an innovative approach for characterizing the volumetrically local ferroelectric and mechanical properties of functional oxides via the application of tomographic Atomic Force Microscopy (T-AFM). Traditional AFM methods are confined to surface studies, whereas our implementation of serial imaging and polishing enables novel through-thickness nanoscale measurements. Specifically for epitaxial bismuth ferrite (BiFeO3) and polycrystalline lead zirconate titanate (PZT) thin films, we investigate the intricate domain structures, and additional complexity associated with grain boundaries. Such research, with depth sensitivity down to the Angstrom scale, is crucial for discovering and optimizing mesoscale, thickness dependent, and even emergent functional properties. These findings have broad applications ranging from energy solutions to beyond-CMOS devices.

CONTROL ID: 3983515
CURRENT SYMPOSIUM: S14 - Emerging Semiconductor Materials and Interfaces
CURRENT SESSION: 2D materials - synthesis, characterization, theory
TITLE: Freestanding Nanomembranes as New Material Building Blocks: From Artificial Heterostructures to Monolithic 3D integration
PRESENTATION TYPE: Invited (by Invitation Only)
PRESENTER (FIRST NAME ONLY): Sanghoon
PRESENTER (LAST NAME ONLY): Bae
PRESENTER (INSTITUTION ONLY): Washington University, St. Louis
PRESENTER (EMAIL ONLY): sbae22@wustl.edu
PRESENTER (COUNTRY ONLY): United States
DATE/TIME SUBMITTED: September 20, 2023, 01:37 PM
ABSTRACT BODY:
Abstract Body: In this talk, I would like to share our team's endeavor to realize next-generation material building blocks, freestanding nanomembranes, and their applications. First of all, we have conceived an approach to obtain freestanding single crystalline materials through 2D materials-assisted layer transfer (2DLT). While developing it, we found out that crystallographic information can penetrate through graphene as long as substrates have polarity because of graphene's information transparency. Also, the slippery nature of graphene enables spontaneous relaxation, which substantially reduces dislocation density in heteroepitaxy. Second, we realized that geometrical confinement facilitates kinetic control of 2D materials, which secures crystallinity, layer-controllability, and heterostructures. Thereby we for the first time succeeded in producing single crystalline 2D materials on amorphous substrates without an epitaxial relationship. Lastly, the development of 3D and 2D nanomembranes enables the creation of a new class of heterostructures, called artificial heterostructures, and vertically integrated devices. These unique structures exhibit exciting physical couplings and hold great promise for practical applications.

CONTROL ID: 3983565
CURRENT SYMPOSIUM: S1: Characterization of Structure-Property Relationships in Functional Ceramics
CURRENT SESSION: Advances in connecting local and global structure to properties
TITLE: Uncovering the Origin of Piezoelectric Response in BFTM-PT
PRESENTATION TYPE: Poster
PRESENTER (FIRST NAME ONLY): Michelle
PRESENTER (LAST NAME ONLY): Dolgos
PRESENTER (INSTITUTION ONLY): University of Calgary
PRESENTER (EMAIL ONLY): michelle.dolgos1@ucalgary.ca
PRESENTER (COUNTRY ONLY): Canada
DATE/TIME SUBMITTED: September 20, 2023, 03:08 PM
ABSTRACT BODY:
Abstract Body: Our current work has found a novel high Curie temperature piezoelectric ceramic (T_c = 630C). This material exists in a solid solution with $Bi(Fe_{2/8}Ti_{3/8}Mg_{3/8})O_3$ and $PbTiO_3$ (BFTM-PT), and exhibits enhanced properties at the MPB (d_{33} = 100 pC/N). The work presented here seeks to understand the mechanism behind the high piezoelectric response at the MPB within the BFTM-PT series. In-situ electric field X-ray diffraction was used to track the change in the average and local structures along the BFTM-PT solid solution upon application of electric field. Piezoelectric ceramics often have a complex structure where their local, short-range structure is different from the average, long-range structure. Consequently, under applied electric field, the local and average structures will behave differently providing insight to the mechanism driving electromechanical properties. Capturing the change in local structure (via PDF) and average structure (via diffraction) under applied electric field can provide insight to the origin of the piezoelectric response.

CONTROL ID: 3983589
CURRENT SYMPOSIUM: S14 - Emerging Semiconductor Materials and Interfaces
CURRENT SESSION: WBG and UWBG materials - synthesis, characterization, theory
TITLE: Evaluating Transitions from Axial to Lateral Growth in GaN Nanowire
PRESENTATION TYPE: Contributed (Oral)
PRESENTER (FIRST NAME ONLY): Sadegh
PRESENTER (LAST NAME ONLY): Yazdi
PRESENTER (INSTITUTION ONLY): University of Colorado, Boulder
PRESENTER (EMAIL ONLY): saya9347@colorado.edu
PRESENTER (COUNTRY ONLY): United States
DATE/TIME SUBMITTED: September 25, 2023, 07:04 PM
ABSTRACT BODY:
Abstract Body: One of the primary challenges impeding the progress of wide and ultra-wide bandgap (UWBG) devices lies in the scarcity of high-quality substrates. Epitaxial lateral overgrowth (ELO) offers a promising solution for growing large, high-quality UWBG substrates. Among the various ELO techniques, the lateral growth and coalescence of nanowires, facilitated by the selective area growth method, have recently garnered attention for their effectiveness in suppressing threading dislocations.

In this presentation, I will share our endeavors, utilizing advanced transmission electron microscopy (TEM) techniques, to explore the interface between axially grown GaN nanowire cores and their laterally overgrown shells. Transitioning from axial to lateral growth involves a significant alteration in growth temperature and the III/V precursor ratio when employing the metal-organic vapor phase epitaxy method.

Our investigations show this change does not introduce dislocations but perhaps generates a considerable number of electrically active point defects. I will share our discovery of unintentionally formed quantum wells when transitioning from axial to lateral growth. Our electron holography and electron energy loss spectroscopy measurements reveal an electrostatic potential change and a bandgap drop at these quantum wells.

These findings address questions from previous studies, enhancing our understanding of UWBG device growth.

CONTROL ID: 3983607
CURRENT SYMPOSIUM: S3 - Advanced Electronic Materials: Processing and Structures and Properties and Applications
CURRENT SESSION: Performance, reliability, and fatigue of ferroelectrics and related devices
TITLE: The Effect of Lead Stoichiometry on the Dielectric and Polarization Properties of PSZT(Nb) Ceramics
PRESENTATION TYPE: Contributed (Oral)
PRESENTER (FIRST NAME ONLY): Sean
PRESENTER (LAST NAME ONLY): Bishop
PRESENTER (INSTITUTION ONLY): Sandia National Laboratories
PRESENTER (EMAIL ONLY): srbisho@sandia.gov
PRESENTER (COUNTRY ONLY): United States
DATE/TIME SUBMITTED: September 25, 2023, 02:41 PM
ABSTRACT BODY:
Abstract Body: This work examines the impact of Pb content on microstructure, phase, dielectric, polarization, and temperature and pressure induced phase transformations of Nb-doped, Sn-modified PZT 95/5 (PZST(Nb)). The Pb content is modified by batching to -4, -2, 0, and 2 mol% Pb excess. Increasing Pb content is found to increase the grain size, reduce ZrO_2 impurity, increase theoretical density, decrease relative density, decrease elastic modulus and Poisson's ratio, decrease rhombohedral low to high phase transition temperature, and increase Curie temperature. Coercive field was unchanged with Pb content, while relative permittivity may decrease with increasing Pb. The temperature dependence of dielectric and polarization behavior is also presented. The relationship between trends with ZrO_2 impurity, tolerance factor, and grain size is discussed.

CONTROL ID: 3983618

CURRENT SYMPOSIUM: S16 - New Frontiers and Science in Additive Manufacturing of Ceramic Materials

CURRENT SESSION: New approaches for AM processes of ceramics and ceramic-matrix composites

TITLE: Additive Manufacturing of Thermochromic Oxide Ceramic Coatings via Inorganic Binder Condensation Using Flash Lamp Annealing

PRESENTATION TYPE: Contributed (Oral)

PRESENTER (FIRST NAME ONLY): Eren

PRESENTER (LAST NAME ONLY): Ozmen

PRESENTER (INSTITUTION ONLY): Georgia Institute of Technology

PRESENTER (EMAIL ONLY): eozmen3@gatech.edu

PRESENTER (COUNTRY ONLY): United States

DATE/TIME SUBMITTED: September 20, 2023, 09:19 PM

ABSTRACT BODY:

Abstract Body: The use of flash lamp annealing(FLA) techniques (that can deliver up to 30 J/cm^2 in a single pulse) to bind and/or sinter alumina coatings with a phosphate binder (dihydrogen phosphate, ADP,$Al(H_2PO_4)_3$)) is studied for additive manufacturing(AM) purposes. Doping alumina powders with chromium makes them useful as thermochromic temperature sensors. The phase evolution and microstructural development of spray case Al_2O_3 and Cr:Al_2O_3 film are studied as a function of FLA heating conditions. As sprayed layers are about 600 mm thick but consolidate to as thin as 45 mm upon heat treatment. Using x-ray diffraction, we can track the transformation of ADP to the aluminum metaphosphate ($AlPO_4$) phase. Minimum FLA energy densities of 450-500 J/cm^2(which take 45 seconds to 1 minute to deliver) are determined to be sufficient to drive the ADP to $AlPO_4$ transformation. Thermochromism of these Cr:Al_2O_3 coatings are tested using a UV-Vis spectrometer and show good thermochromic transitions (color change between pink and gray) even after 10 rounds of cyclic heating and cooling between 400 °C and 25 °C. Notably, the FLA excels in time-saving and its capacity to efficiently cure large surface areas using photonic energy, and it can be offered as an alternative to laser-based methods for producing stable oxide ceramic coatings with a potential for scaling to other AM approaches.

CONTROL ID: 3983676
CURRENT SYMPOSIUM: S4 - Complex Oxide Thin Films and Heterostructures: From Synthesis to Strain/Interface-engineered Emergent Properties
CURRENT SESSION: Phenomena arising from strain couplings and interface couplings including quantum phases and topological states and their control via external stimuli
TITLE: Magnetism in iron oxide perovskites: Synthesis, stability and strain
PRESENTATION TYPE: Invited (by Invitation Only)
PRESENTER (FIRST NAME ONLY): Jennifer
PRESENTER (LAST NAME ONLY): Fowlie
PRESENTER (INSTITUTION ONLY): SLAC National Accelerator Laboratory
PRESENTER (EMAIL ONLY): jfowlie@stanford.edu
PRESENTER (COUNTRY ONLY): Switzerland
DATE/TIME SUBMITTED: September 20, 2023, 08:26 PM
ABSTRACT BODY:
Abstract Body: Solid state systems possessing non-trivial topological properties are of interest in the context of spintronic and memory applications as well as the fundamental study of quantum materials. A particularly fascinating example of such a system is the correlated oxide strontium ferrite, $SrFeO_3$, due to its persistent metallicity and plethora of magnetic phases that arise from the high symmetry of its crystal structure. One type of multi-q helimagnetic structure is suggested to host a skyrmion lattice, an appealing prospect for future integration into oxide electronic devices. With this in mind, there is a clear need for high quality $SrFeO_3$ thin films, typically a challenge due to an unfavourable iron oxidation state and materials degradation. We synthesize and characterize $SrFeO_3$ thin films grown by pulsed laser deposition, addressing these challenges by varying growth conditions, substrates and encapsulating layers.

CONTROL ID: 3983689

CURRENT SYMPOSIUM: S1: Characterization of Structure-Property Relationships in Functional Ceramics

CURRENT SESSION: Advances in connecting local and global structure to properties

TITLE: Practical Magnetic Measurements for Functional Ceramics

PRESENTATION TYPE: Contributed (Oral)

PRESENTER (FIRST NAME ONLY): Eve

PRESENTER (LAST NAME ONLY): Mozur

PRESENTER (INSTITUTION ONLY): Colorado School of Mines

PRESENTER (EMAIL ONLY): evemozur@mines.edu

PRESENTER (COUNTRY ONLY): United States

DATE/TIME SUBMITTED: September 20, 2023, 07:14 PM

ABSTRACT BODY:

Abstract Body: Magnetic materials are useful for a wide variety of applications, from data storage to renewable energy. However, many researchers consider magnetometry and other magnetic characterization methods to be solely useful for traditional magnetic applications. We hypothesize that the gap in applications and use stems from a gap in information between what one can learn in a textbook and what one needs to conduct experimentation. For example, experimental design can be highly sample dependent. Here, we will discuss the practicalities of magnetic measurements to demystify magnetometry for non-experts through illustrative examples.

CONTROL ID: 3983775

CURRENT SYMPOSIUM: S4 - Complex Oxide Thin Films and Heterostructures: From Synthesis to Strain/Interface-engineered Emergent Properties

CURRENT SESSION: Strain- and interface-controlled device performance in energy harvesting, memories, sensors, etc.

TITLE: Effect of Heterointerfaces Formed by Metal Exsolution on the Thermoelectric Properties of Epitaxial $La_{0.7}Ca_{0.2}Ni_{0.25}Ti_{0.75}O_3$ Thin Films

PRESENTATION TYPE: Invited (by Invitation Only)

PRESENTER (FIRST NAME ONLY): Dongkyu

PRESENTER (LAST NAME ONLY): Lee

PRESENTER (INSTITUTION ONLY): University of South Carolina

PRESENTER (EMAIL ONLY): dongkyu@cec.sc.edu

PRESENTER (COUNTRY ONLY): United States

DATE/TIME SUBMITTED: September 24, 2023, 05:48 PM

ABSTRACT BODY:

Abstract Body: Forming a heterointerface by dispersing metallic particles into materials is an effective approach for altering the electrical transport properties. This strategy can be extended to enhance the power factor (PF) of transition metal oxides, where the intrinsic trade-off relationship between the electrical conductivity (σ) and thermopower (S) limits the PF. Here, we report the influence of heterointerfaces, formed by the exsolution of metal nanoparticles from perovskite lattices, on the thermoelectric (TE) properties of epitaxial $La_{0.7}Ca_{0.2}Ti_{0.75}Ni0_{.25}O_3$ (LCTNO) thin films. We create a heterointerface between metal nanoparticles and the oxide through B-site exsolution. Metal exsolution leads to a decrease in the carrier concentration, while increasing the carrier mobility. Consequently, σ decreases, whereas S increases. However, the exsolution of metal particles increases S by up to 20 times while causing only a threefold decrease in σ compared to one before exsolution. Thus, the overall enhancement of PF is by 2 orders of magnitude. We attribute this enhancement to the selectively filtered carriers resulting from the carrier energy band bending formed at the metal-oxide interfaces. This study provides a novel route to tailoring the interconnected physical and chemical properties of oxides, leading to the enhancement in the TE power output.

CONTROL ID: 3983778
CURRENT SYMPOSIUM: S10 - Fundamental mechanisms in materials for next generation computing devices
CURRENT SESSION: Scalable modeling and computation for switching phenomena
TITLE: Deterministic Conductive Filament Dynamics for Improved Switching Uniformity in Embedded Metal-Oxide-Based Memristors
PRESENTATION TYPE: Invited (by Invitation Only)
PRESENTER (FIRST NAME ONLY): Ye
PRESENTER (LAST NAME ONLY): Cao
PRESENTER (INSTITUTION ONLY): University of Texas at Arlington
PRESENTER (EMAIL ONLY): ye.cao@uta.edu
PRESENTER (COUNTRY ONLY): United States
DATE/TIME SUBMITTED: September 21, 2023, 01:47 AM
ABSTRACT BODY:
Abstract Body: Device-to-device variation of switching performance is one of the major obstacles preventing the applications of metal oxide-based memristors in large-scale memory storage and resistive neural networks. Recent experimental works have reported that embedding metal nano-islands (NIs) in metal oxide can effectively improve the uniformity of the memristors, but the underlying role of the NIs is not fully understood. Here, we develop a physical model to understand the origin of the variability and how the embedded NIs can improve the uniformity of memristors. We find that embedding metal NIs can modulate the electric field distribution and lead to a more deterministic formation of the conductive filament (CF) from their vicinity, in contrast to random growth of CFs without embedded NIs. This further enhances the uniformity of the operation voltages and current ON/OFF ratios. We further demonstrate that choosing NI metal composition and shape that chemically facilitate vacancy formations can further optimize the CF morphology, reduce operation voltages, and improve switching performance. Our work provides a fundamental understanding of how embedded metal NIs improve the resistive switching performance in oxide-based memristors and could potentially guide the selection of embedded NIs to realize a more uniform memristor.

CONTROL ID: 3983811
CURRENT SYMPOSIUM: S3 - Advanced Electronic Materials: Processing and Structures and Properties and Applications
CURRENT SESSION: Advanced electronic materials, including ferroelectric, piezoelectric, dielectric, electrostrictive, electrocaloric and pyroelectric materials
TITLE: $Pb(Fe_{0.5}Nb_{0.5})O_3$–$BiFeO_3$ bulk ceramics and screen-printed films for energy storage applications
PRESENTATION TYPE: Contributed (Oral)
PRESENTER (FIRST NAME ONLY): Ivana
PRESENTER (LAST NAME ONLY): Goričan
PRESENTER (INSTITUTION ONLY): Jozef Stefan Institute
PRESENTER (EMAIL ONLY): ivana.gorican@ijs.si
PRESENTER (COUNTRY ONLY): Slovenia
DATE/TIME SUBMITTED: September 25, 2023, 07:36 AM
ABSTRACT BODY:
Abstract Body: Energy storage is the ability to collect and store energy for use at another time. Dielectric capacitors are good candidates for energy storage in pulsed systems. The most promising are dielectric materials with slim and pinched polarization versus electric field (P–E) hysteresis loops. Such properties are obtained in (1–x)$Pb(Fe_{0.5}Nb_{0.5})O_3$–x$BiFeO_3$ (PFN–100xBFO) solid solutions. Compared to bulk ceramics, thick films withstand higher electric fields, which makes them even more promising for energy storage applications.

In this work, we investigated the energy storage properties of PFN–100xBFO bulk ceramics and thick films. PFN–100xBFO (x = 0–0.5) powders were prepared by mechanochemical synthesis, isostatically pressed into pellets and sintered in an oxygen atmosphere at 900 °C or 950 °C for 2 h. The best energy-storage properties were determined for the composition x = 0.3. The recoverable energy storage density (U_{rec}) and efficiency (η) values were 1.0 J/cm^3 and 91 % at 70 kV/cm, respectively. After that, thick films of the composition x = 0.3 were screen-printed on golden Al_2O_3 substrates. The samples were sintered in an oxygen atmosphere at 800 °C for 2 h. The U_{rec} value of the PFN–30BFO thick films reached 2.8 J/cm^3 at 400 kV/cm and the η value was 79 %. The samples were additionally cycled by an electric field and survived 10 million cycles.

CONTROL ID: 3983879
CURRENT SYMPOSIUM: S16 - New Frontiers and Science in Additive Manufacturing of Ceramic Materials
CURRENT SESSION: New approaches for AM processes of ceramics and ceramic-matrix composites
TITLE: Infrared Irradiation to Drive Phosphate Condensation as a Route to Direct Additive Manufacturing of Oxide Ceramics
PRESENTATION TYPE: Contributed (Oral)
PRESENTER (FIRST NAME ONLY): Mark
PRESENTER (LAST NAME ONLY): Losego
PRESENTER (INSTITUTION ONLY): Georgia Institute of Technology
PRESENTER (EMAIL ONLY): losego@gatech.edu
PRESENTER (COUNTRY ONLY): United States
DATE/TIME SUBMITTED: September 21, 2023, 09:40 AM
ABSTRACT BODY:
Abstract Body: This talk will describe a fast, low-temperature, pressureless process to chemically bind ceramic parts via infrared (IR) irradiation and phosphate binder condensation. Ceramic components are synthesized from slurries of ceramic powders and $Al(H_2PO_4)_3$ binder that are irradiated with short-waved IR light capable of heating the system to 350°C. This irradiation is sufficient to drive phosphate condensation, binding the ceramic powders together within a matter of seconds. The IR-irradiated components show an increase in density and Vickers hardness. Layer-by-layer spraying and irradiation is demonstrated as a route to additive manufacturing using various ceramic chemistries including Al_2O_3, Fe_2O_3, TiO_2, In_2O_3 and others. While further optimization is needed to control desired microstructure, this process of using chemically bonded ceramic binders with IR heating for additive manufacturing shows the potential to find applications in various ceramic systems, including refractories, bone implants, electronics, and thermal barrier coatings.

CONTROL ID: 3983890

CURRENT SYMPOSIUM: S4 - Complex Oxide Thin Films and Heterostructures: From Synthesis to Strain/Interface-engineered Emergent Properties

CURRENT SESSION: Controlled synthesis of epitaxial thin films and nanocomposites, including heterostructures, superlattices, and vertically aligned nanocomposites for functional properties

TITLE: Crystallization Kinetics in Atomic Layer Deposited (ALD) Thin Films: Opportunities of Time and Chemical Environment

PRESENTATION TYPE: Contributed (Oral)

PRESENTER (FIRST NAME ONLY): Mark

PRESENTER (LAST NAME ONLY): Losego

PRESENTER (INSTITUTION ONLY): Georgia Institute of Technology

PRESENTER (EMAIL ONLY): losego@gatech.edu

PRESENTER (COUNTRY ONLY): United States

DATE/TIME SUBMITTED: September 21, 2023, 10:09 AM

ABSTRACT BODY:

Abstract Body: Atomic layer deposition (ALD) frequently produces amorphous films, however crystalline phases are often desired for various applications. Understanding the fundamental crystallization kinetics is important in designing process pathways to achieve low-temperature and ultra-thin film (<50 nm) crystallization. This presentation will discuss our work to study the crystallization kinetics in ALD TiO_2 thin films. We focus on TiO_2 ALD via tetrakis(dimethylamino)titanium(IV) (TDMAT) and water. Time-dependent transformation kinetics are fit to the Johnson-Mehl-Avrami-Kolmogorov (JMAK) model, revealing a continuous nucleation process and a 2D growth mechanism, consistent with the observed grain sizes. Images collected at short annealing times are used to independently determine the crystal nucleation rate, allowing us to determine that the activation energies for nucleation (1.32–1.35 eV K^{-1} $atom^{-1}$) are about one order of magnitude greater than the activation energies for grain growth (0.12–0.24 eV K^{-1} $atom^{-1}$), sugggesting that nucleation is the more significant barrier to crystallization in this system. Finally, we show how a modification to the process's chemical environment can lower this barrier to nucleation leading to dramatic in situ crystallization (with no post-deposition annealing) at unusually low process temperatures and film thicknesses (< 15 nm).

CONTROL ID: 3983905
CURRENT SYMPOSIUM: S11 - In situ/operando Characterization of Nanomaterials
CURRENT SESSION: Advances in developing in situ characterization techniques related to electron microscopy, X-ray, and neutron characterization techniques
TITLE: Polarized Neutron Reflectivity Under Pressure and Other Stimuli
PRESENTATION TYPE: Invited (by Invitation Only)
PRESENTER (FIRST NAME ONLY): Timothy
PRESENTER (LAST NAME ONLY): Charlton
PRESENTER (INSTITUTION ONLY): Oak Ridge National Lab
PRESENTER (EMAIL ONLY): charltontr@ornl.gov
PRESENTER (COUNTRY ONLY): United States
DATE/TIME SUBMITTED: September 21, 2023, 10:48 AM
ABSTRACT BODY:
Abstract Body: Polarized neutron reflectivity provides a non-destructive measure of the chemical and magnetic density profile normal to the surface and along the surface projection of the scattering plane. The atomic contrast varies randomly across the periodic table and available isotopes. Isotopic substitution can be used to tune the contrast of a layer to optimize the scattering potential in order to highlight specific features of interest. Sensitivity to magnetism is provided by the interaction of the neutrons spin with the magnetic induction (mu . B). In situ and in operando sample environment equipment design benefit from the lack of absorption for most materials. This allows a significant amount of material along the neutron beam path when required such as in pressure and fluid cells. In this presentation, we will describe a pressure cell designed for neutron reflection studies and initial experimental results. We will also cover examples where a film is excited by microwave pumping, exotic sample design that allows for the application of thermal gradients, and describe the extraction of reflectivity curves when the sample is driven by a high frequency stimulus (i.e. greater than 1kHz).

CONTROL ID: 3983964

CURRENT SYMPOSIUM: S9 - Machine Learning and Automated Synthesis/Characterization for Novel Materials

CURRENT SESSION: Data-driven modeling and analysis

TITLE: Rapid Material Spectra Identification with a Transferable Vision Transformer Model

PRESENTATION TYPE: Contributed (Oral)

PRESENTER (FIRST NAME ONLY): Yunchao

PRESENTER (LAST NAME ONLY): Xie

PRESENTER (INSTITUTION ONLY): University of Missouri

PRESENTER (EMAIL ONLY): ycxie2010@gmail.com

PRESENTER (COUNTRY ONLY): United States

DATE/TIME SUBMITTED: September 21, 2023, 12:46 PM

ABSTRACT BODY:

Abstract Body: Efficiently analyzing material spectra is crucial to prevent the accumulation of unwieldy datasets that could impede decision-making processes. However, existing methodologies heavily rely on manual expertise and domain knowledge, which not only proves laborious but also struggles to keep pace with the rapid data acquisition. In this context, we present a novel approach that leverages a transferable Vision Transformer (ViT) model for precise and expedited identification of materials spectra, including XRD and FTIR. When applied to XRD spectra classification, the optimal ViT model achieves impressive prediction accuracies of 70% (Top-1), 93% (Top-3), and 94.9% (Top-5), while significantly reducing training time to just 269 seconds compared to a convolutional neural network model. Dimension reduction and attention weight maps highlight its proficiency in capturing relevant features within XRD spectra to determine prediction outcomes. Furthermore, the ViT model can seamlessly transfer to FTIR spectra classification, achieving remarkable Top-1, Top-3, and Top-5 prediction accuracies of 84%, 94.1%, and 96.7%, respectively. This study presents an accurate and interpretable approach that accelerates the analysis of materials spectra, providing a unified platform for handling various types of spectroscopic data. Our work paves the way for expediting materials characterization processes.

CONTROL ID: 3984046
CURRENT SYMPOSIUM: S18 - Structure and function in high-entropy, entropy-stabilized, and related oxides
CURRENT SESSION: Understanding and predicting structure and properties through computation and data science
TITLE: Short range ordering in entropy stabilized oxides with the α-PbO_2 structure
PRESENTATION TYPE: Invited (by Invitation Only)
PRESENTER (FIRST NAME ONLY): Solveig
PRESENTER (LAST NAME ONLY): Aamlid
PRESENTER (INSTITUTION ONLY): University of British Columbia
PRESENTER (EMAIL ONLY): solveig.aamlid@ubc.ca
PRESENTER (COUNTRY ONLY): Canada
DATE/TIME SUBMITTED: September 25, 2023, 10:47 PM
ABSTRACT BODY:
Abstract Body: Short range ordering (SRO) is an elusive phenomenon to characterize and quantize, but also potentially pivotal for the emergence of superior functional properties in high entropy materials. In this work, we aim to compare the SRO of two entropy stabilized oxides which both crystallize in the α-PbO_2 structure, the two-component (Ti, Zr)O_2 and the four-component (Ti, Zr, Hf, Sn)O_2. Pair distribution function measurements reveal large changes in the local structure of (Ti, Zr)O_2 depending on quenching versus slow cooling of the sample, while (Ti, Zr, Hf, Sn)O_2 display no visible changes. A computational approach based on density functional theory, cluster expansions, and Metropolis Monte Carlo simulated annealing is used to explain the experimental observations. The order-disorder phase transition temperature in the four-component compound is lowered by hundreds of degrees relative to the two-component compound, ensuring that the sluggish diffusion kinetics keep the cations frozen in place regardless of cooling rate, effectively hindering any SRO. The two-component compound on the other hand, is showing signs of SRO even at synthesis temperature, indicating that the ideal configurational entropy state is never reached.

CONTROL ID: 3984107
CURRENT SYMPOSIUM: S11 - In situ/operando Characterization of Nanomaterials
CURRENT SESSION: Advances in developing in situ characterization techniques related to electron microscopy, X-ray, and neutron characterization techniques
TITLE: Infrared Energy Excitations at High and Low Temperatures with In Situ Monochromated EELS
PRESENTATION TYPE: Invited (by Invitation Only)
PRESENTER (FIRST NAME ONLY): Jordan
PRESENTER (LAST NAME ONLY): Hachtel
PRESENTER (INSTITUTION ONLY): Oak Ridge National Lab
PRESENTER (EMAIL ONLY): Jordan.Hachtel@gmail.com
PRESENTER (COUNTRY ONLY): United States
DATE/TIME SUBMITTED: September 22, 2023, 05:52 PM
ABSTRACT BODY:
Abstract Body: Modern monochromators can achieve energy resolution on the order of a few meV, making a wide range of infrared excitations such as phonons, molecular vibrations, and low-energy plasmons/excitons readily accessible in the electron microscope. As with many systems, in situ heating or cooling experiments enable new aspects of the material to be probed through phase transitions or through how various collective excitaitons react to the presence (or absence) of thermal energy. The ability to pair ultrahigh energy resolution EELS with in situ temperature control is a brand new combined experimental approach with scarce few real examples, but a tremendours amount of potential in the fields of biological, quantum, and photonic systems. Here, I will present some initial results on heating and cooling experiments with monochromated EELS experiments that provide exciting insights into a variety of systems, as well as provide an outlook into the future of in situ monochromated EELS for electronic materials.

CONTROL ID: 3984115
CURRENT SYMPOSIUM: S15 - Electronics Materials for 5G/6G Applications
CURRENT SESSION: 5G/6G Metrology
TITLE: Frequency-Dependent Electromagnetic Properties of Engineered Flexible Composites
PRESENTATION TYPE: Invited (by Invitation Only)
PRESENTER (FIRST NAME ONLY): Luckshitha
PRESENTER (LAST NAME ONLY): Suriyasena Liyanage
PRESENTER (INSTITUTION ONLY): University of Colorado Boulder
PRESENTER (EMAIL ONLY): luckshitha.suriyasenaliyanage@nist.gov
PRESENTER (COUNTRY ONLY): United States
DATE/TIME SUBMITTED: September 25, 2023, 01:40 AM
ABSTRACT BODY:
Abstract Body: Flexible engineered materials are widely being incorporated in wearable and biomedical electronics space. Recent developments in flexible and wearable electronics also incorporate circuits, systems and IoT that communicate at high frequencies (> 1 GHz) using 5G and 6G communication standards. Thus it is imperative for scientists and manufacturers to accurately evaluate the electromagnetic properties of flexible materials that can have applications in 5G/6G millimeter-wave technologies.

We present a novel, flip-chip technique to extract the electromagnetic properties of flexible materials that employs on-wafer error correction and de-embedding. Herein we measure the signal propagation through a CPW transmission line, calibrate out the effects of the probe, and utilize the fringing fields penetrating the material to extract its properties. We engineered flexible composites with magnetic and conductive nanoparticles such as barium hexaferrite (BaM), carbon nanotubes (CNTs), graphite and reduced graphene oxide (rGO) and extracted their magnetic, dielectric and conductive properties to demonstrate this technique. Such composites can be directly utilized in conductive interconnects, flexible antennas, and electromagnetic absorbers. This research encourages and invites the study of a wide array of materials that can have promising applications in flexible electronics and communications industry.

CONTROL ID: 3984121
CURRENT SYMPOSIUM: S17 - Electronic and Ionic Materials in Battery Energy Storage
CURRENT SESSION: Next Gen Electrolytes
TITLE: Processing-Structure-Transport Property Relationships of Pulsed-Laser Deposited LATP as Thin-Film Battery Solid Electrolytes
PRESENTATION TYPE: Contributed (Oral)
PRESENTER (FIRST NAME ONLY): En Ju
PRESENTER (LAST NAME ONLY): Cho
PRESENTER (INSTITUTION ONLY): University of Illinois at Urbana-Champaign
PRESENTER (EMAIL ONLY): enjucho2@illinois.edu
PRESENTER (COUNTRY ONLY): United States
DATE/TIME SUBMITTED: September 25, 2023, 09:23 PM
ABSTRACT BODY:
Abstract Body: Solid-state electrolyte development is important for future battery technologies, with advances in safety and high-power density vs. liquid electrolytes. $Li_{1.3}Al_{0.3}Ti_{1.7}(PO_4)_3$, LATP, exhibits high ionic conductivity and a wide electrochemical stability window, suitable for use with high voltage cathodes. We study the pulsed laser deposition fabrication parameter – structure – conductivity relationships of LATP thin films. We have shown that while ablation of a single LATP target results in Li- and P-deficient films, we can control both the Li and phosphate (PO_4) content by alternating deposition from LATP and Li_3PO_4 targets. To the best of our knowledge, no publications reported the use of Li_3PO_4 as an additional target for multilayer deposition. Not only the chemical compositions, but also the crystallinity and phase composition of films are altered by varying PLD parameters and heat treatments. We therefore study how growth temperature, gas pressure, multilayer ratio, and post-deposition anneals affect the in-plane ionic conductivity of the films with in-situ ac-impedance spectroscopy vs. temperature. Preliminary results show that low PO_2 during deposition causes depletion of Li and PO_4, lowering the conductivity. Our study provides a route for deposition of thin films with multiple volatile species with high ionic conductivity.

CONTROL ID: 3984155
CURRENT SYMPOSIUM: S3 - Advanced Electronic Materials: Processing and Structures and Properties and Applications
CURRENT SESSION: Materials design, new materials and structures, and their emerging applications
TITLE: Optical Furnace for High-Throughput Sintering
PRESENTATION TYPE: Poster
PRESENTER (FIRST NAME ONLY): Jaime
PRESENTER (LAST NAME ONLY): Tellez Gonzalez
PRESENTER (INSTITUTION ONLY): Colorado School of Mines
PRESENTER (EMAIL ONLY): jtellezgonzalez@mines.edu
PRESENTER (COUNTRY ONLY): United States
DATE/TIME SUBMITTED: September 25, 2023, 02:38 PM
ABSTRACT BODY:
Abstract Body: In classic ceramic processing, shaped powder compacts are sintered into a consolidated mass at elevated temperatures. The sintering temperature, dwell time, and heating rate all significantly influence the final microstructure and properties of the ceramic. In conventional sintering, the entire furnace and its insulation must be heated and cooled, which limits the attainable heating rates and reduces efficiency at the laboratory scale. To address the significant time and energy requirements associated with conventional sintering, and to increase the range of attainable microstructures, alternative methods have been developed to reduce sintering times and temperatures. In this work, we developed a small optical image furnace that can rapidly sinter ceramic specimens and anneal them by directly heating them with halogen lamps. The specimen temperature is measured with a thermocouple, and the sample is monitored with a digital video feed. Here we compare conventionally sintered and optically sintered $BaTiO_3$. Optical sintering drastically reduces sintering times from several hours to a few minutes, but large thermal gradients can cause microstructural heterogeneity and potential failures due to thermal stresses. Ongoing work is focused on rapid real-time monitoring and control to address such issues. We demonstrate the optical furnace as a proof-of-concept for potential use in high-throughput sintering of small ceramic specimens.

CONTROL ID: 3984227
CURRENT SYMPOSIUM: S14 - Emerging Semiconductor Materials and Interfaces
CURRENT SESSION: WBG and UWBG materials - synthesis, characterization, theory
TITLE: Photoluminescence mapping of surface defects on β-Ga_2O_3
PRESENTATION TYPE: Contributed (Oral)
PRESENTER (FIRST NAME ONLY): Matthew
PRESENTER (LAST NAME ONLY): McCluskey
PRESENTER (INSTITUTION ONLY): Washington State University
PRESENTER (EMAIL ONLY): mattmcc@wsu.edu
PRESENTER (COUNTRY ONLY): United States
DATE/TIME SUBMITTED: September 22, 2023, 12:21 AM
ABSTRACT BODY:
Abstract Body: Monoclinic gallium oxide (β-Ga_2O_3) is an ultrawide bandgap semiconductor with potential applications in power electronics. Photoluminescence (PL) spectroscopy is an important method to characterize dopants and defects in this material. Common features in the PL spectrum include the intrinsic UV band, blue and green bands that involve donor-acceptor pairs, and red emission due to Cr^{3+} impurities. PL mapping with excitation wavelengths ranging from 266 to 532 nm reveals the spatial distribution of these features with micron resolution. In Czochralski-grown β-Ga_2O_3, the Cr^{3+} emission intensity shows striations that are attributed due to inhomogeneities during growth. In addition to defects in the bulk, PL microscopy has revealed several specific defects on the surface. Some of these localized centers are very bright UV emitters. Homoepitaxial layers show defects that are observed via the shifts in the PL band, likely due to the strain field around a dislocation core. Damage due to high-intensity laser pulses results in significant changes in the intensity and energy of the UV band. In situ PL spectroscopy performed with a pulsed 266 nm laser shows characteristic emission peaks attributed to Ga atoms ablated from the surface.

CONTROL ID: 3984276
CURRENT SYMPOSIUM: S1: Characterization of Structure-Property Relationships in Functional Ceramics
CURRENT SESSION: Advances in scattering, imaging, and analytical techniques
TITLE: On the structures and tunable functionalities in engineered perovskite materials
PRESENTATION TYPE: Invited (by Invitation Only)
PRESENTER (FIRST NAME ONLY): Di
PRESENTER (LAST NAME ONLY): Wang
PRESENTER (INSTITUTION ONLY): Karlsruhe Institute of Technology
PRESENTER (EMAIL ONLY): di.wang@kit.edu
PRESENTER (COUNTRY ONLY): Germany
DATE/TIME SUBMITTED: September 25, 2023, 02:55 AM
ABSTRACT BODY:
Abstract Body: Strongly-correlated phenomena, such as, colossal magnetoresistance (CMR) and metal-insulator transitions (MIT), exhibited by perovskite materials are accompanied and reinforced by coexistence of competing magneto-electronic phases. Such magneto-electronic inhomogeneity is governed by the intrinsic lattice-charge-spin-orbital correlations, which can be tailored via chemical substitution, strain engineering, proximity induced charge redistribution, etc. In this presentation, high resolution scanning transmission electron microscopy (HR-STEM) and integrated differential phase contrast (iDPC) imaging have been used to determine the crystal structure and atomically resolved electron energy-loss spectroscopy (EELS) has revealed the charge transfer and local valence state for two types of engineered perovskite materials: 1) high entropy oxides (HEOs) $(Gd_{0.25}La_{0.25}Nd_{0.25}Sm_{0.25})_{1-x}Sr_xMnO_3$ (x = 0 – 0.5), in which the presence of multiple-principal cations are stabilized on a given sub-lattice, exhibit indications of an inherent magneto-electronic phase separation encapsulated in single crystallographic phase; 2) Sr iridate thin films with proximity induced change transfer from $LaCoO_3$ epitaxially grown on it, showing ferromagnetic state. The atomically resolved geometric and electronic structures are successfully correlated to the finely tuned electromagnetic properties of the investigated materials.

CONTROL ID: 3984294
CURRENT SYMPOSIUM: S5 - Controlling Grain Boundary Structure, Chemistry, and their Network as a Function of Mtl Processing
CURRENT SESSION: Defect and segregation behavior
TITLE: The Consequences of Space-Charge Zones for Short-Circuit Diffusion along Extended Defects
PRESENTATION TYPE: Invited (by Invitation Only)
PRESENTER (FIRST NAME ONLY): Roger
PRESENTER (LAST NAME ONLY): De Souza
PRESENTER (INSTITUTION ONLY): RWTH Aachen University
PRESENTER (EMAIL ONLY): desouza@pc.rwth-aachen.de
PRESENTER (COUNTRY ONLY): Germany
DATE/TIME SUBMITTED: September 22, 2023, 05:31 AM
ABSTRACT BODY:
Abstract Body: For metallic systems, faster diffusion along extended defects, such as grain boundaries and dislocations, is relatively well characterised and understood. This is not the case, however, for oxides, principally because they are far more complex systems. One of the reasons for the higher complexity is that extended defects in oxides are not constrained in equilibrium to remain locally electroneutral. Because of re-distribution of pont defects between an extended defect and the bulk phase, the extended defect will in general become electrostatically charged, with the adjacent bulk phase concomitantly developing diffuse, enveloping space-charge zones.
The altered concentrations of point defects in these space-charge zones have long been understood to affect, for instance, the transport of charge across grain boundaries in polycrystalline systems. In this presentation, I will draw attention to one particular aspect that has received far less attention: that space-charge zones can provide the dominant contribution to short-circuit diffusion along grain boundaries and dislocations. I will also show that space-charge zones can affect the excess Gibbs free energy of a grain boundary. Lastly, I will emphasise the need to consider space-charge zones from a thermodynamic perspective; and that concentrated solid solutions cannot be treated with the standard Poisson–Boltzmann approach.

CONTROL ID: 3984301
CURRENT SYMPOSIUM: S2 - Advanced Microelectronics
CURRENT SESSION: Dielectrics and ferroelectrics for device applications
TITLE: Epitaxial growth of lithium niobate thin films for acoustic resonators in wireless network
PRESENTATION TYPE: Contributed (Oral)
PRESENTER (FIRST NAME ONLY): Zhen
PRESENTER (LAST NAME ONLY): Ye
PRESENTER (INSTITUTION ONLY): Institute of Materials Research and Engineering, A*STAR, Singapore
PRESENTER (EMAIL ONLY): e0816501@u.nus.edu
PRESENTER (COUNTRY ONLY): Singapore
DATE/TIME SUBMITTED: September 22, 2023, 04:33 AM
ABSTRACT BODY:
Abstract Body: Radio frequency (RF) filters based on acoustic resonators are essential components in wireless networks, responsible for selecting the desired frequency bands and filtering out unwanted interference. Aluminium nitride thin films are the dominating material in RF filters for decades, but its low piezoelectric property is difficult to meet the growing demands of wide bandwidth. Lithium niobate ($LiNbO_3$, LN) is a material with a high electromechanical coupling coefficient that enables higher operating frequencies and wider bandwidths. However, the current smart-cut process for fabricating LN thin films is a major obstacle to commercialization due to its complexity and cost. Direct growth of LN thin films is a more scalable approach, but it remains challenging to grow high quality films .

In this research, we propose a modified sputtering method to address the challenges in synthesizing high-quality LN thin films. We focus on two main challenges: achieving stoichiometric composition without generating secondary phases and optimizing the growth parameters to reduce island growth and obtain a smooth surface. We investigate the impact of substrate, gas pressure, gas ratio, and temperature on the crystal quality and surface roughness of LN thin films. These directly grown LN thin films are promising for acoustic resonator and RF filters with wide bandwidth for high-speed wireless networks.

CONTROL ID: 3984317
CURRENT SYMPOSIUM: S7 - Defects and Transport in Ceramics
CURRENT SESSION: Defect mobility and transport behavior, including under operando or in extreme environments
TITLE: Oxygen Surface Exchange Coefficient of $SrTiO_3$: The Effect of Surface Orientation and Water
PRESENTATION TYPE: Invited (by Invitation Only)
PRESENTER (FIRST NAME ONLY): Roger
PRESENTER (LAST NAME ONLY): De Souza
PRESENTER (INSTITUTION ONLY): RWTH Aachen University
PRESENTER (EMAIL ONLY): desouza@pc.rwth-aachen.de
PRESENTER (COUNTRY ONLY): Germany
DATE/TIME SUBMITTED: September 22, 2023, 06:13 AM
ABSTRACT BODY:
Abstract Body: By virtue of their heterogeneous nature, reactions comprising the exchange of a chemical component between a gaseous phase and a crystalline compound are characteristically sluggish. In some cases, the chemical component needs to be extracted from a strongly bonded gaseous species, and this will slow down such reactions even further. A simple example of such a reaction is the exchange of oxygen between dioxygen molecules and a solid oxide. Despite intense work on this topic, debates have raged, and still rage, as to the mechanism of the oxygen exchange reaction and its rate-determining step.

In this contribution, I will present our studies of oxygen isotope exchange on single-crystal $SrTiO_3$, focusing on the effect of surface termination and the effect of H_2O. I will also emphasise the importance of considering charged surfaces, at which the concentrations of charge carriers (the reaction partners of molecules in the gas phase) are not related in any simple manner to charge-carrier concentrations in the bulk phase.

CONTROL ID: 3984367

CURRENT SYMPOSIUM: S11 - In situ/operando Characterization of Nanomaterials

CURRENT SESSION: Study of nanomaterial-based catalysts at reactive environments

TITLE: Dynamics of Nanoparticle Exsolution and Growth: An in-situ STEM study

PRESENTATION TYPE: Contributed (Oral)

PRESENTER (FIRST NAME ONLY): Dylan

PRESENTER (LAST NAME ONLY): Jennings

PRESENTER (INSTITUTION ONLY): Forschungszentrum Juelich

PRESENTER (EMAIL ONLY): d.jennings@fz-juelich.de

PRESENTER (COUNTRY ONLY): Germany

DATE/TIME SUBMITTED: October 08, 2023, 05:58 PM

ABSTRACT BODY:

Abstract Body: Metal exsolution is a novel synthesis route for the formation of nanostructured oxide-supported metal catalysts. Among other benefits, the method allows for the formation of well dispersed and highly active metal nanoparticles. One of the major claims of exsolution as a synthesis process is that an enhanced particle stability is achieved as compared to conventional oxide-supported catalyst synthesis methods, leading to catalytic systems with increased efficiency and extended lifetimes. To fully understand the mechanisms of the increased stability of exsolved nanoparticles, an exploration of the dynamic processes which govern the growth and morphological changes of these particles is critical. In this work, the dynamics of exsolved Ni nanoparticles are investigated using an environmental STEM equipped with a secondary electron detector, providing a route for atomic-resolution characterization of the sample surface during and after the exsolution process. Several dynamic processes are highlighted and characterized by in situ STEM, with particular focus on the observation and characterization of particle growth by both Ostwald ripening and particle migration. Additionally, the impacts of surface space charge and of subsurface defect structures on particle stability are analyzed, and the variations in particle morphology under changing oxygen partial pressures are characterized.

CONTROL ID: 3984370
CURRENT SYMPOSIUM: S14 - Emerging Semiconductor Materials and Interfaces
CURRENT SESSION: 2D materials - synthesis, characterization, theory
TITLE: High Throughput Automated Synthesis and Characterization of Metal Halide Perovskites
PRESENTATION TYPE: Invited (by Invitation Only)
PRESENTER (FIRST NAME ONLY): Mahshid
PRESENTER (LAST NAME ONLY): Ahmadi
PRESENTER (INSTITUTION ONLY): University of Tennessee, Knoxville
PRESENTER (EMAIL ONLY): mahmadi3@utk.edu
PRESENTER (COUNTRY ONLY): United States
DATE/TIME SUBMITTED: September 22, 2023, 11:13 AM
ABSTRACT BODY:
Abstract Body: Metal halide perovskites have garnered considerable attention in the field of optoelectronics due to their exceptional properties. However, so far little has been understood based on the fundamental principles for designing the functional perovskites, which is now crucially decelerating the lab-to-fab transformation and realization of the scalable manufacturing of these materials for optoelectronics. I will discuss the potential of machine learning-driven high throughput automated experiments to expedite the discovery of hybrid perovskites, optimize processing pathways, and enhance the understanding of formation kinetics [1-5]. The workflow demonstrated in this study serves as a powerful tool for constructing detailed chemical maps of perovskite nanocrystal synthesis, enabling tailored customization of their functionalities. Additionally, another study showcases how high throughput automated synthesis provides a comprehensive guide for designing optimal precursor stoichiometry to achieve functional quasi-2D perovskite phases in thin films capable of realizing high-performance optoelectronics [3,4].
References:
1. Yang, J., Ahmadi, M. Nat. Synth (2023).
2. Yang J., et al., MRS Bulletin 48, 164–172 (2023).
3. Yang J, et. al., ChemRxiv (2023).DOI: 10.26434/chemrxiv-2023-x7sfr
4. Yang J, et. al., ChemRxiv (2023). DOI: 10.26434/chemrxiv-2023-zcvl0
5. Sanchez S. L., et. al., Matter, 1–19 (2023).

CONTROL ID: 3984399

CURRENT SYMPOSIUM: S13 - Frontiers in Ferroic oxides: Synthesis, Properties, and Applications

CURRENT SESSION: Advanced synthesis and characterization techniques for ferroic oxides

TITLE: Exploring free-standing $PbZr_{0.2}Ti_{0.8}O_3$ membrane and its nonvolatile gating effect in two-dimensional MoS_2

PRESENTATION TYPE: Contributed (Oral)

PRESENTER (FIRST NAME ONLY): Qiuchen

PRESENTER (LAST NAME ONLY): Wu

PRESENTER (INSTITUTION ONLY): University of Nebraska-Lincoln

PRESENTER (EMAIL ONLY): wuqiuchen@huskers.unl.edu

PRESENTER (COUNTRY ONLY): United States

DATE/TIME SUBMITTED: September 23, 2023, 02:23 AM

ABSTRACT BODY:

Abstract Body: Ferroelectric oxide membranes are promising materials for constructing flexible, low power nanoelectronics. We study the effect of electrodes on ferroelectricity of single crystalline $PbZr_{0.2}Ti_{0.8}O_3$ (PZT) and PZT/$La_{0.67}Sr_{0.33}MnO_3$ (LSMO) membranes by preparing them on different base-layers, including metallic Au, correlated oxide LSMO, and 2D semiconductor MoS_2. Polarization switching in PZT yields nonvolatile modulation of the MoS_2 conduction, with current on/off ratio of up to 2×10^5 at 300 K and no apparent decay for more than 3 days. Piezoresponse force microscopy studies show that the coercive field E_c for the membranes depends sensitively on the base layer and exhibits strong polarization asymmetry. The PZT/LSMO membranes transferred on LSMO show the lowest E_c of about -0.26/+0.28 MV cm^{-1}, smaller than that of epitaxial PZT films. The domain wall (DW) roughness exponent ζ points to 2D random bond disorder dominated DW roughening (ζ = 0.31) at 300 K. Upon thermal quench at progressively higher temperatures, ζ for PZT membranes on Au and LSMO approaches the theoretical value for 1D random bond disorder (ζ = 2/3), while samples on MoS_2 exhibits thermal roughening (ζ = 1/2). The PZT membranes on Au, LSMO, and MoS_2 show T_C values well exceeding the bulk value. Our study provides important material parameters for the functional design of PZT-based flexible nanoelectronics.

CONTROL ID: 3984437
CURRENT SYMPOSIUM: S12 - Magnetic and Superconducting Materials: From Basic Science to Applications
CURRENT SESSION: Other magnetic materials
TITLE: The Role of Particle Size and Composition in Ferrite/Polymer Microwave Absorbers
PRESENTATION TYPE: Poster
PRESENTER (FIRST NAME ONLY): Andrew
PRESENTER (LAST NAME ONLY): Padgett
PRESENTER (INSTITUTION ONLY): Sandia National Laboratories
PRESENTER (EMAIL ONLY): aspadge@sandia.gov
PRESENTER (COUNTRY ONLY): United States
DATE/TIME SUBMITTED: September 22, 2023, 11:35 AM
ABSTRACT BODY:
Abstract Body: As 5G technology becomes ubiquitous and the internet of things expands, sources of microwave EMI are rapidly increasing, and the threat of coupling EM energy into susceptible electronic circuits. Mitigating this threat requires compact, lightweight broadband absorbers to protect sensitive electronics. Magnetite, MnZn ferrite, and NiZn ferrite, alone and in conjunction, have the potential to meet the energy density and EM attenuation performance needs. However, optimizing the absorption loss requires improved understanding of the role of particle size and volume fraction on the complex permittivity and permeability of these materials. This presentation discusses experimental investigation of the particle size and volume fraction on the absorption losses of these ferrites. Complex permeability and permittivity measurements were conducted to identify the absorption mechanisms. We present a possible explanation of these results in the context of Snoek's limit and the Landau-Lifshitz theory of permeability.
This work was supported by the Laboratory Directed Research and Development program at Sandia National Laboratories, a multimission laboratory managed and operated by National Technology and Engineering Solutions of Sandia LLC, a wholly owned subsidiary of Honeywell International Inc. for the U.S. Department of Energy's National Nuclear Security Administration under contract DE-NA0003525.

CONTROL ID: 3984496
CURRENT SYMPOSIUM: S12 - Magnetic and Superconducting Materials: From Basic Science to Applications
CURRENT SESSION: Other magnetic materials
TITLE: Extraordinary and Surface Phase Transitions in a van der Waals Layered Antiferromagnet
PRESENTATION TYPE: Invited (by Invitation Only)
PRESENTER (FIRST NAME ONLY): Xiaoyu
PRESENTER (LAST NAME ONLY): Guo
PRESENTER (INSTITUTION ONLY): University of Michigan
PRESENTER (EMAIL ONLY): guoxy@umich.edu
PRESENTER (COUNTRY ONLY): United States
DATE/TIME SUBMITTED: September 22, 2023, 12:36 PM
ABSTRACT BODY:
Abstract Body: Surface is present in any finite-size material systems, and its impact on phase transitions can be quite unconventional and interesting. Theory predicted a few decades ago that surface can undergo a phase transition into an ordered state at a higher temperature than bulk does when the interaction strength at the surface exceeds that inside the bulk – named as surface and extraordinary phase transitions for surface and bulk, respectively. Van der Waals (vdW) layered magnets provide a unique platform for potentially realizing such a stringent and rare condition. In this presentation, we use a combination of second order nonlinear optics, transport, and transmission electron microscopy to investigate CrSBr, a vdW layered antiferromagnet. We report our finding of surface and extraordinary phase transitions in CrSBr that are separated by 8K.

CONTROL ID: 3984514
CURRENT SYMPOSIUM: S15 - Electronics Materials for 5G/6G Applications
CURRENT SESSION: 5G/6G Devices
TITLE: Leveraging Emerging Magnetic Materials for 6G Devices
PRESENTATION TYPE: Invited (by Invitation Only)
PRESENTER (FIRST NAME ONLY): Piotr
PRESENTER (LAST NAME ONLY): Kulik
PRESENTER (INSTITUTION ONLY): University of Central Florida
PRESENTER (EMAIL ONLY): piotr.kulik@ucf.edu
PRESENTER (COUNTRY ONLY): United States
DATE/TIME SUBMITTED: September 25, 2023, 01:30 PM
ABSTRACT BODY:
Abstract Body: The congestion of the electromagnetic spectrum caused by the ever-increasing use of devices has pushed silicon-based components to their limit. As a result, magnetic thin-films such as YIG (yttrium iron garnet) have gathered attention for their unique properties ranging from non-reciprocity to high frequency operation enabling full-duplex systems and applications in the mm-Wave range. In this talk, we will discuss the challenges and recent progress in the fabrication and integration of such thin-films in rf and optical devices. Furthermore, design of components such as antennas and isolators based on spinwaves will be presented.

CONTROL ID: 3984521
CURRENT SYMPOSIUM: S18 - Structure and function in high-entropy, entropy-stabilized, and related oxides
CURRENT SESSION: ESOs in energy storage
TITLE: Hierarchical Length-Scale Structure Effects of Cation Substitution in Lithium-Excess Disordered Rocksalt Oxides
PRESENTATION TYPE: Contributed (Oral)
PRESENTER (FIRST NAME ONLY): John
PRESENTER (LAST NAME ONLY): Langhout
PRESENTER (INSTITUTION ONLY): University of Florida
PRESENTER (EMAIL ONLY): johnlanghout@ufl.edu
PRESENTER (COUNTRY ONLY): United States
DATE/TIME SUBMITTED: September 22, 2023, 02:45 PM
ABSTRACT BODY:
Abstract Body: Lithium-excess disordered rocksalt oxides (DRXs) have garnered significant attention as potential next-generation lithium-ion battery cathodes, owed to their high energy density and chemical flexibility. Typical DRXs contain two transition metals (TMs) and an excess of Li, meaning they are better classified as medium-entropy materials due to the low number of non-equimolar species. As a result, in contrast to high-entropy materials, the DRX's structure is heavily influenced by the type and concentration of the species within. Here, we study a compositional series of DRXs which gradually replace Ti^{4+} with Zr^{4+}. Resulting changes in the structure at various length-scales are reported using synchrotron X-ray diffraction, pair distribution function analysis, and X-ray absorption spectroscopy. Lattice parameter and nanodomain size, short-range order, and local atomic distortions are all modulated by the concentration of Zr^{4+}, indicating the critical role of each transition metal's size, charge, and electronic properties on the structure, and thus, properties, of this emerging class of materials.

CONTROL ID: 3984537
CURRENT SYMPOSIUM: S4 - Complex Oxide Thin Films and Heterostructures: From Synthesis to Strain/Interface-engineered Emergent Properties
CURRENT SESSION: Characterizations of strain, defects, and interfaces
TITLE: Exploring Oxygen Motion in Hafnium Zirconium Oxide Thin Films for Nonvolatile Memory Applications
PRESENTATION TYPE: Contributed (Oral)
PRESENTER (FIRST NAME ONLY): Liron
PRESENTER (LAST NAME ONLY): Shvilberg
PRESENTER (INSTITUTION ONLY): University of Virginia
PRESENTER (EMAIL ONLY): ls3qt@virginia.edu
PRESENTER (COUNTRY ONLY): United States
DATE/TIME SUBMITTED: September 22, 2023, 04:06 PM
ABSTRACT BODY:
Abstract Body: Since the discovery of the ferroelectric phase in hafnium oxide (HfO_2) based thin films, substantial resources have been drawn to prepare reliable HfO_2-based devices, mainly for nonvolatile memory applications. HfO_2-based thin films, in particular hafnium-zirconium oxide (HZO) films, are studied extensively due to their compatibility with complementary metal oxide semiconductor devices. One of the stabilizing factors of ferroelectric hafnia is oxygen vacancies; however, when accumulated at interfaces it can be detrimental to device performance. Many unclear phenomena witnessed in HZO films, including wake-up, fatigue, and imprint, are challenges that can be traced, at least in part, to oxygen vacancies. To deepen the understanding of what drives these effects, it is crucial to understand the mechanisms controlling oxygen vacancy motion in hafnia-based films and to better tailor the position of the oxygen vacancies throughout the film to improve electrical performance reliability. In this talk the mechanisms of oxygen motion during the processing steps of HZO films and field cycling will be discussed. To study how oxygen diffuses in the HZO film during processing, monitoring oxygen motion in the HZO has been characterized using grazing incident x-ray diffraction, time of flight secondary ion mass spectroscopy, and electrical testing, all for different processing and use conditions.

CONTROL ID: 3984550
CURRENT SYMPOSIUM: S14 - Emerging Semiconductor Materials and Interfaces
CURRENT SESSION: 2D materials - synthesis, characterization, theory
TITLE: Hybrid 2D Van der Waals Materials for Functional Electronic Devices
PRESENTATION TYPE: Invited (by Invitation Only)
PRESENTER (FIRST NAME ONLY): Yu
PRESENTER (LAST NAME ONLY): Zhong
PRESENTER (INSTITUTION ONLY): Cornell University
PRESENTER (EMAIL ONLY): yz2833@cornell.edu
PRESENTER (COUNTRY ONLY): United States
DATE/TIME SUBMITTED: September 22, 2023, 01:50 PM
ABSTRACT BODY:
Abstract Body: Artificial van der Waals (vdW) solids assembled from 2D crystals offer a new frontier in materials engineering at the atomic scale. While extensively explored in inorganic 2D materials for optoelectronic and quantum devices, this talk introduces a new approach: the synthesis of hybrid organic-inorganic vdW heterostructures for electronic applications. We introduce monolayer 2D polymers as emerging 2D building blocks for making vdW solids that offer unlimited structural tunability on the molecular level. Through programmed assembly, we create hybrid 2D heterostructures, seamlessly integrating 2D polymers with inorganic counterparts. This approach allows for integrating versatile molecular building blocks and inorganic monolayers into layered nanomaterials to develop ultrathin electronic and electrochemical devices.

CONTROL ID: 3984597

CURRENT SYMPOSIUM: S17 - Electronic and Ionic Materials in Battery Energy Storage

CURRENT SESSION: Solid State Ion Conductors

TITLE: Interrogating the Effects of Misorientation on the Ion Conductivity of Single Grain Boundaries in Epitaxially Grown $Li_{3x}La_{1/3-x}TaO_3$ Films

PRESENTATION TYPE: Contributed (Oral)

PRESENTER (FIRST NAME ONLY): Ian

PRESENTER (LAST NAME ONLY): Brummel

PRESENTER (INSTITUTION ONLY): University of Virginia

PRESENTER (EMAIL ONLY): iab5dy@virginia.edu

PRESENTER (COUNTRY ONLY): United States

DATE/TIME SUBMITTED: September 25, 2023, 02:22 PM

ABSTRACT BODY:

Abstract Body: Grain boundaries (GBs) are generally highly resistive to ionic transport in solid ion conductors, which is a key challenge impeding widespread use of all solid state lithium ion batteries. While this has been shown repeatedly in the literature using impedance spectroscopy measurements of polycrystalline materials, the resulting measurement only gives an average value for all boundaries in the conduction path. This work presents a novel method for the measurement of single GB behavior as a function of misorientation by employing epitaxial growth of $Li_{3x}La_{1/3-x}TaO_3$ on $SrTiO_3$ bicrystal substrates to generate a well-defined, single GB. By applying photolithograpically defined electrodes, impedance spectroscopy can be measured across a single GB. Understanding of the change in ionic resistivity of grain boundaries as a function of the GB misorientation could provide a straightforward route toward the design of polycrystalline ion conducting materials for battery applications. The epitaxial relationship across the GB is verified using X-ray diffraction. A combination of electrical characterization through impedance spectroscopy and scanning probe microscopy with structural and compositional characterization allow a deep understanding of how the GB structure generates a barrier to ion conduction.

CONTROL ID: 3984637
CURRENT SYMPOSIUM: S5 - Controlling Grain Boundary Structure, Chemistry, and their Network as a Function of Mtl Processing
CURRENT SESSION: Nanocrystalline ceramics
TITLE: FIB-Based TEM Specimen Preparation for Characterizing Nanoscale Microstructures of Ceramics
PRESENTATION TYPE: Invited (by Invitation Only)
PRESENTER (FIRST NAME ONLY): Sarshad
PRESENTER (LAST NAME ONLY): Rommel
PRESENTER (INSTITUTION ONLY): University of Connecticut
PRESENTER (EMAIL ONLY): sarshad.rommel@uconn.edu
PRESENTER (COUNTRY ONLY): United States
DATE/TIME SUBMITTED: September 22, 2023, 08:48 PM
ABSTRACT BODY:
Abstract Body: Transparent ceramics like magnesium aluminate spinels (MAS) are excellent candidates for replacing traditional glasses due to their superior mechanical properties and optical transparency. In nanocrystalline (NC) MAS with extremely small grain sizes, the dominant deformation mechanisms are often grain-boundary-assisted, such as boundary sliding and shear banding. Thus, a detailed characterization of features like grain boundaries and porosity is essential to understand the mechanical properties of NC MAS. Focused ion beam (FIB)- based transmission electron microscopy (TEM) specimen preparation is usually suitable for microstructural characterization of such materials due to constrains on size/quantity of bulk material available. However, standard FIB-based TEM lift-out specimens are too thick (80-120nm) for characterization of nanoscale grains, as grain overlap and interference hinders imaging of features. A protocol has been developed for producing FIB-based TEM specimens for imaging NC MAS nanoscale microstructures (3.7-80nm grain size). In this protocol, after standard lift-out procedure, the thinned lamella is tilted by 2.5-4°, and vertical channels ~1μm wide are thinned at 5kV operating voltage, such that the thickness at lower edge of lamella reaches zero inside the channels. This allows for successful imaging of microstructural features, even for grain sizes as small as 3.7nm.

CONTROL ID: 3984646
CURRENT SYMPOSIUM: S14 - Emerging Semiconductor Materials and Interfaces
CURRENT SESSION: WBG and UWBG materials - synthesis, characterization, theory
TITLE: Atomic-scale insights into ultra-wide bandgap semiconductors for enhanced power electronics
PRESENTATION TYPE: Invited (by Invitation Only)
PRESENTER (FIRST NAME ONLY): Baishakhi
PRESENTER (LAST NAME ONLY): Mazumder
PRESENTER (INSTITUTION ONLY): University at Buffalo - The State University of New York, University at Buffalo - The State University of New York, Buffalo, NY, US, academic
PRESENTER (EMAIL ONLY): baishakh@buffalo.edu
PRESENTER (COUNTRY ONLY): United States
DATE/TIME SUBMITTED: September 24, 2023, 08:28 PM
ABSTRACT BODY:
Abstract Body: Ultra-wide bandgap (UWBG) semiconductors provide a revolutionary solution to surpass silicon technology limitations, especially in high-power and high-temperature electronics. Advancing UWBG semiconductor technology relies on precise material engineering for tailored electrical properties while minimizing defects. This requires a deep understanding of structural and chemical attributes influencing these properties. The design of optimal UWBG-based heterojunction devices necessitates subnanometer-scale characterization. Atom probe tomography (APT) stands out as a premier technique for atomic-scale material analysis enabling comprehensive 3D elemental characterization, encompassing chemical distributions, grain boundaries, interfaces, and dopant profiles. Our research endeavors leverage the capabilities of APT for atomic-scale material characterization, augmented by cutting-edge material informatics-based data analysis methods. Through the application of these techniques, our objective is to uncover the atomic-scale structural and chemical attributes that hinder the optimal performance of UWBG materials. This presentation serves as an enlightening exploration of the indispensable role of atomic-scale analysis in advancing oxide and nitride-based UWBG semiconductors, ultimately contributing to the realization of more efficient, resilient, and high-performance power electronic devices.

CONTROL ID: 3984690
CURRENT SYMPOSIUM: S10 - Fundamental mechanisms in materials for next generation computing devices
CURRENT SESSION: Co-design of materials, devices, and systems
TITLE: Transforming Nanoelectronics with Volatile Memristors
PRESENTATION TYPE: Invited (by Invitation Only)
PRESENTER (FIRST NAME ONLY): Ahmedullah
PRESENTER (LAST NAME ONLY): Aziz
PRESENTER (INSTITUTION ONLY): University of Tennessee Knoxville
PRESENTER (EMAIL ONLY): aziz@utk.edu
PRESENTER (COUNTRY ONLY): United States
DATE/TIME SUBMITTED: September 25, 2023, 09:32 PM
ABSTRACT BODY:
Abstract Body: Relentless miniaturization of transistors, driven by Moore's Law, has long propelled electronics' evolution towards atomic scales. Consequently, quantum phenomena and fundamental physical limits have emerged as pressing concerns in device and circuit design. Traditional CMOS-based electronics are in transition, ushering in the post-CMOS era and sparking a surge in the exploration of novel materials and devices. Volatile memristors, known for their electrically induced phase transitions, stand out as one of the most promising materials with the potential to revolutionize the next generation of electronics. This presentation will showcase several innovative device and circuit concepts that leverage selective resistance switching within these materials. The first segment of the discussion will delve into memristor-augmented spintronic memory devices. This approach capitalizes on the non-volatility of magnetic tunnel junctions, coupled with the abrupt switching and high selectivity of Mott memristors, addressing critical challenges in spin-based memories. The talk will illustrate how this concept extends to memory platforms, arrays, and peripheral circuits, highlighting its transformative potential. The next part of the talk will shift the spotlight onto the oscillatory behavior of volatile Mott memristors and explore their applications in energy-efficient neuromorphic computing systems.

CONTROL ID: 3984727
CURRENT SYMPOSIUM: S13 - Frontiers in Ferroic oxides: Synthesis, Properties, and Applications
CURRENT SESSION: Ferroic materials for novel devices
TITLE: Interface Charge Engineering in Ferroelectric-Gated Mott Transistors
PRESENTATION TYPE: Contributed (Oral)
PRESENTER (FIRST NAME ONLY): Yifei
PRESENTER (LAST NAME ONLY): Hao
PRESENTER (INSTITUTION ONLY): University of Nebraska - Lincoln
PRESENTER (EMAIL ONLY): yfhao@huskers.unl.edu
PRESENTER (COUNTRY ONLY): United States
DATE/TIME SUBMITTED: September 24, 2023, 09:30 PM
ABSTRACT BODY:
Abstract Body: Epitaxial complex oxide heterostructures composed of ferroelectric gates and correlated channels are promising material platforms for developing high performance, energy efficient electronic devices. In this study, working with a ferroelectric $Pb(Zr,Ti)O_3$ gate, we present a systematic study of the ferroelectric field effect in rare earth nickletes $Sm_{0.5}Nd_{0.5}NiO_3$, $NdNiO_3$, and $LaNiO_3$ (LNO) to identify the optimal channel materials for constructing nonvolatile Mott transistors. For single layer nickelate channels, the resistance switching ratio $\Delta R/R_{on}$ peaks near the electrical dead layer thickness, then decreases abruptly due to strong depolarization. In the single layer LNO channels, the ΔR/Ron of 1,619% is observed in 0.8 nm channel thickness due to its smallest electrical dead layer thickness. We also show that inserting a $La_{0.67}Sr_{0.33}MnO_3$ (LSMO) buffer layer can lead to over two orders of magnitude increase in $\Delta R/R_{on}$ for devices with the same channel thickness, which is attributed to the tailored carrier density profile due to the interfacial charge-transfer between nickelate and LSMO. A record high $\Delta R/R_{on}$ of 38,540% is observed in the 1 nm LNO/0.8 nm LSMO bilayer channel at room temperature. Our study addresses one of the key material challenges that limit the application potential of epitaxial ferroelectric-gated Mott transistors.

CONTROL ID: 3984733
CURRENT SYMPOSIUM: S10 - Fundamental mechanisms in materials for next generation computing devices
CURRENT SESSION: Fundamental materials needs for emerging computing paradigms
TITLE: Heat transfer-based nanoscale neuristor
PRESENTATION TYPE: Invited (by Invitation Only)
PRESENTER (FIRST NAME ONLY): Ivan
PRESENTER (LAST NAME ONLY): Schuller
PRESENTER (INSTITUTION ONLY): University of California San Diego
PRESENTER (EMAIL ONLY): ischuller@ucsd.edu
PRESENTER (COUNTRY ONLY): United States
DATE/TIME SUBMITTED: October 06, 2023, 09:03 PM
ABSTRACT BODY:
Abstract Body: Machine learning imitates the basic features of biological neural networks to efficiently perform tasks such as pattern recognition. This has been mostly achieved at a software level, and a strong effort is currently being made to mimic neurons and synapses with hardware components, an approach known as neuromorphic computing. CMOS-based circuits have been used for this purpose, but they are non-scalable, limiting the device density and motivating the search for "neuromorphic materials". While recent advances in resistive switching have provided a path to emulate synapses at the 10 nm scale, a scalable neuron analogue is yet to be found. We developed a heat transfer based Mott nanodevice able to mimic neuron functionalities. We showed that thermal dynamics allow the implementation of the basic neuron functionalities: "leaky integrate-and-fire", refractory period and rate coding. By using local temperature as the internal variable, we avoid the need of external capacitors, which reduces neuristor size by several orders of magnitude. This approach could enable neuromorphic hardware to fully take advantage of the rapid advances in memristive synapses, allowing for much denser and complex neural networks. More generally, we show that heat generation is not always an undesirable effect: it can perform computing tasks if properly engineered.
Supported by the AFOSR under award # FA9550-22-1-0135

CONTROL ID: 3984743
CURRENT SYMPOSIUM: S5 - Controlling Grain Boundary Structure, Chemistry, and their Network as a Function of Mtl Processing
CURRENT SESSION: Nanocrystalline ceramics
TITLE: Mobile dislocation mediated Hall-Petch and inverse Hall-Petch behaviors in nanocrystalline Al-doped boron carbide
PRESENTATION TYPE: Invited (by Invitation Only)
PRESENTER (FIRST NAME ONLY): Qi
PRESENTER (LAST NAME ONLY): An
PRESENTER (INSTITUTION ONLY): Iowa State University
PRESENTER (EMAIL ONLY): qan@iastate.edu
PRESENTER (COUNTRY ONLY): United States
DATE/TIME SUBMITTED: September 22, 2023, 06:32 PM
ABSTRACT BODY:
Abstract Body: In nanocrystalline ceramics, the dynamics behaviors related to the Hall-Petch and inverse Hall-Petch relationships remain elusive, primarily due to the limited dislocation activities, a characteristic feature of metals. Our study employs molecular dynamics simulations with a machine-learning force field to investigate the shear deformation behaviors of nanocrystalline Al-doped boron carbide (n-B12-CAlC) across different grain sizes. We observe an important transition from the Hall-Petch to the inverse Hall-Petch behavior when the grain size reaches a critical 6.09 nm, alongside a peak shear strength of 14.93 GPa. A notable discovery is the activation of mobile dislocations originating from grain boundaries (GBs) in n-B12-CAlC, attributed to the fracture of fragile C-Al chain bonds. This phenomenon significantly influences Hall-Petch behaviors. As grain sizes decrease, enhanced GB regions equalize the shear stress, inhibiting dislocation nucleation from GBs, thus reinforcing the material, in line with the traditional Hall-Petch relationship. Yet, a further reduction in grain size (below approx. 6 nm) offers more opportune sites for dislocation nucleation through the expanded GB regions, leading to a decline in shear strength and initiating the inverse Hall-Petch behavior.

CONTROL ID: 3984752
CURRENT SYMPOSIUM: S10 - Fundamental mechanisms in materials for next generation computing devices
CURRENT SESSION: Fundamental materials needs for emerging computing paradigms
TITLE: High-precision analog computing with memristors
PRESENTATION TYPE: Invited (by Invitation Only)
PRESENTER (FIRST NAME ONLY): Joshua
PRESENTER (LAST NAME ONLY): Yang
PRESENTER (INSTITUTION ONLY): USC
PRESENTER (EMAIL ONLY): jjoshuay@usc.edu
PRESENTER (COUNTRY ONLY): United States
DATE/TIME SUBMITTED: September 22, 2023, 06:45 PM
ABSTRACT BODY:
Abstract Body: In the era of the Internet of Things (IoT), the proliferation of ubiquitous sensors has led to a significant influx of analog data, commonly referred to as the "analog data deluge." This surge necessitates the development of a novel computing paradigm capable of efficiently processing analog data with high throughput. In-memory computing utilizing analog memristors has emerged as a promising solution for this purpose. However, existing efforts have primarily focused on low-precision domains like neural networks or have relied on frequent quantization, leading to notable energy and area efficiency drawbacks. To enhance the precision of analog computing, recent advancements have been made on two fronts. Firstly, memristors with an unprecedented precision of thousands of conductance levels have been developed, surpassing the precision of all other analog memory types. Secondly, a novel circuit architecture and programming protocol have been designed to achieve remarkably high precision while minimizing circuit overhead. These breakthroughs enable analog devices to perform high-precision computing, significantly expanding the range of applications for analog computing.

CONTROL ID: 3984803
CURRENT SYMPOSIUM: S13 - Frontiers in Ferroic oxides: Synthesis, Properties, and Applications
CURRENT SESSION: Spin-orbit electronics
TITLE: Sputtering Growth of Nanometer-Thick $Y_3Fe_5O_{12}$ Films
PRESENTATION TYPE: Invited (by Invitation Only)
PRESENTER (FIRST NAME ONLY): Mingzhong
PRESENTER (LAST NAME ONLY): Wu
PRESENTER (INSTITUTION ONLY): Northeastern University
PRESENTER (EMAIL ONLY): min.wu@northeastern.edu
PRESENTER (COUNTRY ONLY): United States
DATE/TIME SUBMITTED: September 24, 2023, 04:29 PM
ABSTRACT BODY:
Abstract Body: Magnetic damping in yttrium iron garnet (YIG) $Y_3Fe_5O_{12}$ is lower than in any other magnetic materials. As such, YIG materials have been widely used in microwave devices. Recent years witnessed a strong interest in the use of YIG materials for spintronic and magnonic device applications, but such.applications require YIG films with a thickness in the nanometer (nm) range. This presentation reports (1) the feasibility of using magnetron sputtering to grow nm-thick YIG films with very low damping, (2) sputtering growth of YIG thin films with both low damping and perpendicular magnetic anisotropy, (3) patterning of YIG thin films with e-beam lithography and dry etching, and (4) fabrication of YIG/topological insulator bi-layered structures.

CONTROL ID: 3984826
CURRENT SYMPOSIUM: S13 - Frontiers in Ferroic oxides: Synthesis, Properties, and Applications
CURRENT SESSION: Electrical control of magnetism and spin textures
TITLE: Magnetic structure in Colossal Magneto resistance in pure and Ge-doped$Mn_3Si_2Te_6$
PRESENTATION TYPE: Invited (by Invitation Only)
PRESENTER (FIRST NAME ONLY): Feng
PRESENTER (LAST NAME ONLY): Ye
PRESENTER (INSTITUTION ONLY): Oak Ridge National Lab
PRESENTER (EMAIL ONLY): yefengcn@gmail.com
PRESENTER (COUNTRY ONLY): United States
DATE/TIME SUBMITTED: September 22, 2023, 10:54 PM
ABSTRACT BODY:
Abstract Body: Recently discovered colossal magnetoresistance (CMR) $Mn_3Si_2Te_6$ has shown stark contrast with conventional CMR materials. The resistivity in $Mn_3Si_2Te_6$ drops by orders of magnitude leading to an insulator-metal transition with an applied magnetic field just above 4 T. The CMR occurs only when the field is applied along the hard c-axis, while such effect is absent with the field applied in the basal plane, where magnetization is fully saturated. We have conducted systematic single crystal neutron diffraction study of the ground state spin structures of pure and Ge-doped $Mn_3Si_2Te_6$ and the evolution of the spin configurations as function of temperature, electric current, and external magnetic field up to 14 Tesla along the c-axis. Prominent short-range magnetic diffuse scattering near Tc, the close connection between the square of spin-spin correlation lengths and electric resistivity, strongly emphasizes the relevance of fluctuating moments contributing to the transport properties. Application of magnetic field along the c axis renders a swift occurrence of CMR but only a slow tilting of the magnetic moments toward the c axis. The results provide direct evidence that the modification in spin order is not sufficient to explain the giant magnetoresistance response when the field is applied along the spin hard axis and reveals new mechanism to describe the novel magnetotransport behavior.

CONTROL ID: 3984844
CURRENT SYMPOSIUM: S9 - Machine Learning and Automated Synthesis/Characterization for Novel Materials
CURRENT SESSION: Autonomous research
TITLE: From automated to autonomous synthesis of inorganic thin film materials
PRESENTATION TYPE: Invited (by Invitation Only)
PRESENTER (FIRST NAME ONLY): Andriy
PRESENTER (LAST NAME ONLY): Zakutayev
PRESENTER (INSTITUTION ONLY): National Renewable Energy Laboratory
PRESENTER (EMAIL ONLY): andriy.zakutayev@nrel.gov
PRESENTER (COUNTRY ONLY): United States
DATE/TIME SUBMITTED: September 22, 2023, 11:43 PM
ABSTRACT BODY:
Abstract Body: Automation of thin film materials synthesis and characterization instruments over the past two decades has led to increased throughout of discovery and design of inorganic materials [1]. In turn these experimental advanced necessitated electronic research data infrastructure [2] and enabled publicly available repositories of the resulting chemical, structural, and synthesis data [3].
The next challenge is how to automate the process of converting data into knowledge and hence select the synthesis and characterization conditions at which the following experiments should be performed. Addressing this challenge will enable fully autonomous experimental laboratories, that take input from previous experiments and make algorithmic decisions on which next experiment to do.
In this presentation, I will describe the transition from automated to autonomous synthesis and characterization experimental equipment at NREL. Examples will include sputter deposition of nitride thin film for photoelectrochemical applications with material composition controlled by Bayesian optimization of optical plasma emission [4], and molecular beam epitaxy of ultra wide band gap oxide semiconductors controlled with feedback from in-situ electron diffraction measurements.
[1] Applied Physics Reviews 4, 011105 (2017)
[2] Patterns 2 100373 (2021)
[3] Scientific data, 5 180053 (2018)
[4] APL Materials 11, 071119 (2023)

CONTROL ID: 3984855

CURRENT SYMPOSIUM: S16 - New Frontiers and Science in Additive Manufacturing of Ceramic Materials

CURRENT SESSION: In-situ and ex-situ characterization of phase transformation, textures, defects, etc., using synchrotron X-ray, neutron diffraction, electron microscopy

TITLE: In situ and ex situ studies of densification issues in ceramic additive manufacturing

PRESENTATION TYPE: Invited (by Invitation Only)

PRESENTER (FIRST NAME ONLY): Andrew

PRESENTER (LAST NAME ONLY): Allen

PRESENTER (INSTITUTION ONLY): NIST

PRESENTER (EMAIL ONLY): andrew.allen@nist.gov

PRESENTER (COUNTRY ONLY): United States

DATE/TIME SUBMITTED: September 23, 2023, 12:45 AM

ABSTRACT BODY:

Abstract Body: While ceramics additive manufacturing (AM) or 3D printing could transform the ceramics industry by allowing complex and customized ceramic components to be fabricated efficiently with minimal waste and environmental impacts, many fundamental challenges remain. Multiple complex materials phenomena are involved. These range from particle–particle interactions that govern feedstock rheology, through formation of various intrinsic defect structures during 3D printing, itself, to densification issues that arise during post-processing. All these phenomena need to be better understood and controlled to achieve high-quality dense AM ceramic parts that perform reliably, e.g., in 4-point bend strength tests. This paper will report progress in addressing these issues, especially for the direct-ink-write ceramic AM route [1], through combined feedstock characterization, in situ rheological studies, in situ and ex situ X-ray and neutron structure/microstructure characterization, and computational modelling. Insights will also be presented from recent in situ X-ray measurements on one of the candidate post-build densification routes via cold sintering [2].

[1] A.J. Allen, I. Levin & R.A. Maier. Int. J. Ceram. Eng. Sci., 4, 302-308 (2022).

[2] F. Zhang et al. Acta Mater., 259, art. no. 119283 (2023).

CONTROL ID: 3984957
CURRENT SYMPOSIUM: S12 - Magnetic and Superconducting Materials: From Basic Science to Applications
CURRENT SESSION: Other magnetic materials
TITLE: Spin-mechanical coupling and dynamics in strained magnetic membranes
PRESENTATION TYPE: Invited (by Invitation Only)
PRESENTER (FIRST NAME ONLY): Jun
PRESENTER (LAST NAME ONLY): Xiao
PRESENTER (INSTITUTION ONLY): University of Wisconsin-Madison
PRESENTER (EMAIL ONLY): jun.xiao@wisc.edu
PRESENTER (COUNTRY ONLY): United States
DATE/TIME SUBMITTED: September 24, 2023, 12:28 PM
ABSTRACT BODY:
Abstract Body: The interactions among spin, charge, and lattice are expected to be substantially enhanced due to the weak dielectric screening at the 2D limit. This enhancement can lead to various exotic spin orderings, including collinear magnets, noncollinear spin textures, quantum spin liquid, and magnetic topological insulators. Additionally, the large mechanical flexibility of magnetic membranes allows unprecedented strain engineering. In this talk, I will present our studies of spin-mechanical coupling and dynamics in strained Heusler compound membranes and vdW layered materials. For the former, spin orderings and carrier dynamics show a strong dependence on the local strain gradient distrubitions probed by magneto-optical microscopy and ultrafast spectroscopy. I will then uncover the spin-mechanical coupling is bidirectional, where magnetization changes can also affect mechanical properties. We interrogate the magnetostriction response in vdW magnets using nano-optomechanical resonators. The developed optomechanical resonators are sensed by precise optical interferometry, whose response shows distinct magnetostrictions in different spin orderings. Moreover, the magnetostriction is tunable by gate-induced strain. Our findings advance the understanding of the coupling of light, magnetism, and mechanics at the nanoscale in emergent magnetic membranes for potential information storage and transduction applications.

CONTROL ID: 3984985
CURRENT SYMPOSIUM: S9 - Machine Learning and Automated Synthesis/Characterization for Novel Materials
CURRENT SESSION: Autonomous research
TITLE: Autonomous Atomic Layer Synthesis via Combinatorial Pulsed Laser Deposition
PRESENTATION TYPE: Invited (by Invitation Only)
PRESENTER (FIRST NAME ONLY): Ichiro
PRESENTER (LAST NAME ONLY): Takeuchi
PRESENTER (INSTITUTION ONLY): University of Maryland
PRESENTER (EMAIL ONLY): takeuchi@umd.edu
PRESENTER (COUNTRY ONLY): United States
DATE/TIME SUBMITTED: September 23, 2023, 02:57 PM
ABSTRACT BODY:
Abstract Body: We are incorporating active learning in combinatorial exploration of functional materials in autonomous modes. Autonomous experimentation can be used to reduce the number of required experimental cycles by an order of magnitude or more. The array format with which samples of different compositions are laid out on combinatorial libraries is particularly conducive to active learning. Using the closed-loop autonomous materials exploration and optimization (CAMEO) scheme, we have discovered a new phase change memory material which outperforms the industrial standard by a significant margin. I will discuss our lastest autonomous experimentation projects including real-time experiment-theory closed-loop interaction and autonomous live control of laser molecular beam epitaxy of functional oxide thin films. This work is performed in collaboration with M. Lippmaa, H. Liang, Y. Sun, A. McDannald, H. Yu, J.-C. Zhao, and A. G. Kusne. This work is funded by NIST, SRC, ONR, ICAM, and AFOSR.

CONTROL ID: 3984999
CURRENT SYMPOSIUM: S13 - Frontiers in Ferroic oxides: Synthesis, Properties, and Applications
CURRENT SESSION: Spin-orbit electronics
TITLE: Low Loss Spinel Ferrite Thin Films for Spin Wave Spintronics
PRESENTATION TYPE: Invited (by Invitation Only)
PRESENTER (FIRST NAME ONLY): Yuri
PRESENTER (LAST NAME ONLY): Suzuki
PRESENTER (INSTITUTION ONLY): Stanford University
PRESENTER (EMAIL ONLY): ysuzuki1@stanford.edu
PRESENTER (COUNTRY ONLY): United States
DATE/TIME SUBMITTED: September 23, 2023, 04:47 PM
ABSTRACT BODY:
Abstract Body: Spin-wave-based electronics based on ferromagnetic insulators where information can be transported and manipulated without charge current would provide a new paradigm for low energy consumption computing. However, insulating behavior is not a sufficient requirement for low damping. We have developed a new class of low loss spin ferrite thin films with Gilbert damping parameter as low as a ~ 0.0006 and negligible inhomogeneous linewidth broadening, resulting in narrow half-width half-maximum linewidths. The most promising spinel ferrite films are of the compositions $Mg(Al,Fe)_2O_4$ and $Li_{0.5}(Al,Fe)_{2.5}O_4$. We have also demonstrated efficient spin pumping from these spinel ferrites into an adjacent heavy metal layer through measurement of the spin-mixing conductance, Gilbert damping enhancement and electrical voltage peaks that appear at ferromagnetic resonance. Spin-torque ferromagnetic resonance measurements indicate that we can indeed achieve electrical control of magnetization. We have also demonstrated current induced spin-orbit torque switching of the ferrite magnetization with spin-to-charge interconversion in Pt that is significantly larger than previous studies. This new class of epitaxial spinel ferrite materials is promising for future low-power electronics such as spin wave logic devices, voltage controlled magnetic memory and magnonic waveguides.

CONTROL ID: 3985030
CURRENT SYMPOSIUM: S13 - Frontiers in Ferroic oxides: Synthesis, Properties, and Applications
CURRENT SESSION: Low dimension ferroic oxides
TITLE: Nanoscale Influences on polar phase transitions in bulk materials
PRESENTATION TYPE: Invited (by Invitation Only)
PRESENTER (FIRST NAME ONLY): Andrew
PRESENTER (LAST NAME ONLY): Rappe
PRESENTER (INSTITUTION ONLY): University of Pennsylvania
PRESENTER (EMAIL ONLY): rappe@sas.upenn.edu
PRESENTER (COUNTRY ONLY): United States
DATE/TIME SUBMITTED: September 23, 2023, 06:35 PM
ABSTRACT BODY:
Abstract Body: The ferroelectric phase transition is generally studied as a bulk phenomenon, but finite-size effects play a considerable role in enhancing or suppressing desired behaviors. Here, we present experimental and theoretical findings characterizing how finite size elicits new polar phenomena. First, we report theoretical and experimental evidence that ordered superlattices can foster a transition from conventional ferroelectric order to relaxor ferroelectricity, with no cation disorder. Second, we study the suppression of relaxor behavior (e.g. dielectric dispersion) when a classic relaxor ferroelectric material's (PMN-PT) thickness is reduced to the nanoscale. Finally, we report a new method for modeling the coupled electric and magnetic polarizations at the million-atom scale in multiferroic BiFeO3. This method can have wide applicability for understanding coupled switching responses in nanoscale multi-domain polar materials.

CONTROL ID: 3985109

CURRENT SYMPOSIUM: S3 - Advanced Electronic Materials: Processing and Structures and Properties and Applications

CURRENT SESSION: Advanced electronic materials, including ferroelectric, piezoelectric, dielectric, electrostrictive, electrocaloric and pyroelectric materials

TITLE: Preparation of functional composite films using the aerosol deposition method

PRESENTATION TYPE: Contributed (Oral)

PRESENTER (FIRST NAME ONLY): Hana

PRESENTER (LAST NAME ONLY): Uršič

PRESENTER (INSTITUTION ONLY): Jozef Stefan Institute

PRESENTER (EMAIL ONLY): hana.ursic@ijs.si

PRESENTER (COUNTRY ONLY): Slovenia

DATE/TIME SUBMITTED: September 24, 2023, 07:34 AM

ABSTRACT BODY:

Abstract Body: In this contribution, we investigate, the feasibility of preparing composite thick films by aerosol deposition (AD). We have started our work by preparing AD ceramic-metal composites from aluminum and alumina Al_2O_3 powders, as these powders have been shown to have high AD efficiencies. In this way, we have demonstrated that AD is an appropriate technique for the preparation of multilayer and composite structures. We have continued our research by preparing multicaloric composites. The electrocaloric (EC) material $0.9Pb(Mg_{1/3}Nb_{2/3})O_3$–$0.1PbTiO_3$ (PMN–10PT) has a large EC temperature change ΔT_{EC}~3.5 °C at 160 kV/cm and a wide working temperature range of ~120 °C. On the other hand, a La–Fe–Si-based magnetocaloric (MC) alloy shows MC temperature change ΔT_{MC} ~2.5 °C at 1.5 T and near room temperature, and a temperature range of ~20 °C. These two materials were used for the preparation of composites with different MC concentrations, spanning from 2 wt% to 20 wt%. The films were prepared with N_2 as the carrier gas on a flexible gold-sputtered polymer substrate and annealed at 400 °C. The 10 wt% films showed a saturation polarization of 36 μC/cm^2 at 600 kV/cm, while the PMN–10PT films showed the same polarization at 900 kV/cm. This enhancement in polarization is due to the Maxwell-Wagner effect. During the presentation, the caloric properties of the composites will be discussed.

CONTROL ID: 3985120
CURRENT SYMPOSIUM: S2 - Advanced Microelectronics
CURRENT SESSION: Emergent devices for novel computing such as neuromorphic computing
TITLE: Harnessing Variability of Ferroelectric and Ferromagnetic Neuromorphic Devices
PRESENTATION TYPE: Contributed (Oral)
PRESENTER (FIRST NAME ONLY): A N M Nafiul
PRESENTER (LAST NAME ONLY): Islam
PRESENTER (INSTITUTION ONLY): Pennsylvania State University
PRESENTER (EMAIL ONLY): nafiul@psu.edu
PRESENTER (COUNTRY ONLY): United States
DATE/TIME SUBMITTED: September 24, 2023, 11:11 AM
ABSTRACT BODY:
Abstract Body: Aspiring to replicate the biological brain's computational efficacy and efficiency, neuromorphic computing has emerged as a disruptive paradigm for future datacentric applications. The one-to-one mapping with brain primitives, such as synapses and neurons, afforded by emerging nanoelectric devices lies at the core of this promise. Wide-scale practical adoption, however, has been limited due to intrinsic device non-idealities and device-to-device variations. Furthermore, scaling down leads to probabilistic device operation. The brain is innately stochastic and thus alludes to the possibility of harnessing device variations and stochasticity for computing. Here, we look at recent advancements in understanding and exploiting device non-idealities for neuromorphic applications using emerging ferro-devices. Ferroelectric and Ferromagnetic devices are at the forefront of beyond CMOS technologies owing to their low power and fast operation, excellent endurance, and compatibility with existing CMOS electronics. We explore how scaling impacts device properties and how inherent variations in device dynamics can lead to novel algorithms – highlighting the need for hardware-software co-design. Moreover, compensation of device non-idealities through in-hardware training is demonstrated, showcasing the potential of large-scale neuromorphic hardware with ferro-devices to achieve truly autonomous edge-intelligence.

CONTROL ID: 3985128

CURRENT SYMPOSIUM: S18 - Structure and function in high-entropy, entropy-stabilized, and related oxides

CURRENT SESSION: Characterization and quantification of structure and properties

TITLE: $Mg_{1/5}Co_{1/5}Ni_{1/5}Cu_{1/5}Zn_{1/5}O$: Growth Kinetics and Nanostructures

PRESENTATION TYPE: Contributed (Oral)

PRESENTER (FIRST NAME ONLY): Saeed

PRESENTER (LAST NAME ONLY): Almishal

PRESENTER (INSTITUTION ONLY): Penn State

PRESENTER (EMAIL ONLY): ssa5409@psu.edu

PRESENTER (COUNTRY ONLY): United States

DATE/TIME SUBMITTED: September 25, 2023, 08:45 PM

ABSTRACT BODY:

Abstract Body: Since Rost et al.'s pivotal report in 2015 on $Mg_{1/5}Co_{1/5}Ni_{1/5}Cu_{1/5}Zn_{1/5}O$ (commonly referred to as J14) [1], HEOs have garnered significant interest for their profound chemical flexibility and local crystalline disorder. The inherent metastability of HEOs results in strong kinetic dependencies on thermal history. The non-equilibrium kinetics of pulsed laser deposition (PLD) rapidly condenses precursors from a high-entropy initial state, allowing kinetic stabilization of a broader spectrum of atomic and electronic configurations than bulk synthesis under near-equilibrium conditions. Presently, we show that epitaxial thin films of J14 accommodate large amount of epitaxial strain and reveal a systematic trend in lattice parameters and properties as a function of substrate temperature and film growth rate. Slow growth rates or elevated growth temperatures facilitate a controllable degree of phase decomposition and subtle nanoscale microstructural reconfiguration during epitaxial growth, including spinel nanocubes and Cu-rich tweed structures embedded in a rocksalt matrix. The morphological evolution of these nanostructures is captured by phase-field modeling. Since physical properties depend on local chemistry and symmetry, controlling the kinetic pathways available to HEOs provides new opportunities for property tuning and development.

[1] Rost et al. "Entropy-stabilized oxides." Nature communications 6.1 (2015): 8485.

CONTROL ID: 3985136
CURRENT SYMPOSIUM: S5 - Controlling Grain Boundary Structure, Chemistry, and their Network as a Function of Mtl Processing
CURRENT SESSION: Defect and segregation behavior
TITLE: Reading between the grains of oxide-ion conductors: The thermal history of space-charge layers
PRESENTATION TYPE: Contributed (Oral)
PRESENTER (FIRST NAME ONLY): Adrian
PRESENTER (LAST NAME ONLY): Usler
PRESENTER (INSTITUTION ONLY): RWTH Aachen University
PRESENTER (EMAIL ONLY): usler@pc.rwth-aachen.de
PRESENTER (COUNTRY ONLY): Germany
DATE/TIME SUBMITTED: September 25, 2023, 01:40 PM
ABSTRACT BODY:
Abstract Body: The excess grain-boundary resistivity in oxide-ion conducting polycrystals is generally attributed to the presence of space-charge layers in which oxygen vacancies are depleted. A simple description of the grain-boundary impedance with closed-form expressions can be provided if the acceptor cations take a uniform concentration profile throughout the sample (Mott–Schottky case). Analysing impedance data with these formulas, however, is problematic in two aspects: first, they are invalid for heavily substituted electrolyte materials, the conductivities of which are known to be largely determined by point-defect interactions. Second, ceramic samples are usually prepared by sintering at high temperatures, thus permitting the acceptor cations to attain electrochemical equilibrium before their movement is frozen upon cooling.

In this work, we present impedance spectra calculated from continuum-level simulations of space-charge layers at a grain boundary. Simulations are carried out in the scope of the restricted-equilibrium and the Poisson–Cahn space-charge models to account for thermal history and for point-defect interactions, respectively. We discuss the errors that arise from an analysis of these more complex cases with the simplified Mott–Schottky model and propose a consistency check to reveal cases that are not adequately described by the Mott–Schottky model.

CONTROL ID: 3985154
CURRENT SYMPOSIUM: S4 - Complex Oxide Thin Films and Heterostructures: From Synthesis to Strain/Interface-engineered Emergent Properties
CURRENT SESSION: Theory, modeling and first principles calculations of complex oxides and their heterostructures
TITLE: You don't have to call them complexions, just don't call them phases
PRESENTATION TYPE: Contributed (Oral)
PRESENTER (FIRST NAME ONLY): W. Craig
PRESENTER (LAST NAME ONLY): Carter
PRESENTER (INSTITUTION ONLY): Massachusetts Institute of Technology
PRESENTER (EMAIL ONLY): wcraigcarter@gmail.com
PRESENTER (COUNTRY ONLY): United States
DATE/TIME SUBMITTED: September 24, 2023, 01:12 PM
ABSTRACT BODY:
Abstract Body: Ming Tang, Rowland Cannon, and I coined the name "Complexions" in 2006[*] The initial motivation was to draw a parallel with wetting transitions and Cahn's critical wetting theory. Our intention was to simplify the description of transitions at grain boundaries between states of differing structure or chemistry. Using "complexions" avoids calling such interfacial chemistries and structures "phases"---which can lead to confusion and misapplication of thermodynamics.

The introduction of the word complexions was surprisingly controversial. Some found it a useful. Others abhorred it.

In this talk, I will illustrate the distinction between a complexion transition and a phase transition. I will review the simple thermodynamics framework for their description and how to relate experimental observations to thermodynamic quantities.

[*] Diffuse interface model for structural transitions of grain boundaries, Ming Tang, W Craig Carter, Rowland M Cannon, Phys. Rev. B, (2006) https://link.aps.org/doi/10.1103/PhysRevB.73.024102

CONTROL ID: 3085182
CURRENT SYMPOSIUM: S11 - In situ/operando Characterization of Nanomaterials
CURRENT SESSION: Study of nanomaterial-based catalysts at reactive environments
TITLE: Breaking the Resolution-Field of View Impasse in Catalyst Characterization with 4D-STEM
PRESENTATION TYPE: Invited (by Invitation Only)

PRESENTER (FIRST NAME ONLY): Debangshu
PRESENTER (LAST NAME ONLY): Mukherjee
PRESENTER (INSTITUTION ONLY): Oak Ridge National Lab
PRESENTER (EMAIL ONLY): mukherjeed@ornl.gov
PRESENTER (COUNTRY ONLY): United States

DATE/TIME SUBMITTED: September 24, 2023, 03:20 PM

ABSTRACT BODY:

Abstract Body: The interplay between microstructure and catalyst behavior is pivotal as catalyst activity often hinges on deviations from the ideal bulk structure, such as lattice strain. Yet, quantifying these deviations with atomic precision remains challenging, given that catalytic activity primarily occurs at the surface. Transmission Electron Microscopy (TEM) has been the tool of choice for materials scientists, offering sub-angstrom precision in quantifying atom positions.
However, specimens observed in TEM for such quantification typically involves a field of view of some tens of nanometers – thus providing information that may not be representative of the entire sample. Here, in this talk I will demonstrate how we can solve this with 4D-STEM (Four-Dimensional Scanning TEM) – where we can obtain high precision strain information across hundreds of particles. This involves collecting 4D-STEM in the nanobeam mode and developing analysis routines to automatically quantify strain from all the particles imaged.
In the second half of this talk, I will also introduce results with electron ptychography, where the 4D-STEM data, when collected with defocused probes, can image large fields of view with sub-angstrom resolution. Ptychographic data additionally retains 3D information, and we will demonstrate how such data can be used for large area tomography of two-dimensional transition metal dichalcogenides.

CONTROL ID: 3985185
CURRENT SYMPOSIUM: S12 - Magnetic and Superconducting Materials: From Basic Science to Applications
CURRENT SESSION: Other magnetic materials
TITLE: Dynamically tunable, narrowband terahertz emitters designed by tailoring magnon-phonon interaction in multilayer heterostructures
PRESENTATION TYPE: Invited (by Invitation Only)
PRESENTER (FIRST NAME ONLY): Jiamian
PRESENTER (LAST NAME ONLY): Hu
PRESENTER (INSTITUTION ONLY): University of Wisconsin-Madison
PRESENTER (EMAIL ONLY): jhu238@wisc.edu
PRESENTER (COUNTRY ONLY): United States
DATE/TIME SUBMITTED: September 24, 2023, 03:12 PM
ABSTRACT BODY:
Abstract Body: A dynamically tunable, narrowband terahertz (THz) source is much needed for an accurate identification and imaging of chemical and biological species, with wide-ranging applications from security screening to medical imaging and to semiconductor inspection. However, existing pulsed THz sources are mostly single-cycle, broadband, and difficult to tune dynamically. In this presentation, the speaker will first introduce an approach that permits converting a femtosecond laser pulse into a multi-cycle, narrowband THz pulse via resonant magnon-phonon interaction. The peak frequency of the emitted THz pulse can be tuned by up to 50% through a varying bias magnetic field. The speaker will then present computational demonstration of this approach in several carefully designed ferromagnet and antiferromagnet based multilayer heterostructures, which allows us to cover without gaps different frequency ranges (deep millimeter-wave, sub-THz, and >1 THz) of the THz spectrum, tune the polarization, duration, and power of the emitted THz pulse. The demonstration was performed based on an in-house dynamical phase-field model that incorporates the coupled dynamics of magnons, phonons, photons, and plasmons in multiphase systems.

CONTROL ID: 3985241
CURRENT SYMPOSIUM: S11 - In situ/operando Characterization of Nanomaterials
CURRENT SESSION: In operando study of device performances and behaviors including batteries, fuel cells, solar cells, memristors, memory devices, etc.
TITLE: In situ Electrical Characterization of Non-volatile Memory Devices under Mixed Radiation Environment
PRESENTATION TYPE: Contributed (Oral)
PRESENTER (FIRST NAME ONLY): Chaitanya
PRESENTER (LAST NAME ONLY): Sharma
PRESENTER (INSTITUTION ONLY): University of Florida
PRESENTER (EMAIL ONLY): chaitanya.sharma@ufl.edu
PRESENTER (COUNTRY ONLY): United States
DATE/TIME SUBMITTED: September 25, 2023, 02:18 PM
ABSTRACT BODY:
Abstract Body: Computational demands expressed in floating-point operations per second (FLOPs) have doubled every three months in the last decade, much faster than Moore's law. Novel non-volatile memory (NVM) architectures stand out as one of the most compelling avenues for addressing the von Neumann bottleneck. However, the effect of in-service radiation exposure from cosmic rays, solar storms, and secondary sources is relatively unexplored. To address this void, we have developed a testing platform for in situ electrical characterization under simultaneous or standalone gamma and neutron radiation environments, with a total flux of 2.035 x 10^8 particles/cm^2. Located inside the south beam port of the 100kW thermal nuclear reactor at the University of Florida training reactor (UFTR), we showcase the versatility of the platform through measurements on commercially available self-directed channel (SDC) memristors under varying reactor power levels across a range from 1W to 10kW. In addition, we will present a comparison between pre-, post-, and operando performance, including current-voltage characteristics, endurance, and retention in the other resistive switching- and ferroelectric-based devices. Degradation and failure modes during operation will be discussed.

CONTROL ID: 3985243

CURRENT SYMPOSIUM: S4 - Complex Oxide Thin Films and Heterostructures: From Synthesis to Strain/Interface-engineered Emergent Properties

CURRENT SESSION: Controlled synthesis of epitaxial thin films and nanocomposites, including heterostructures, superlattices, and vertically aligned nanocomposites for functional properties

TITLE: Epitaxial stabilization of strontium germanates thin films by suboxide molecular-beam epitaxy

PRESENTATION TYPE: Invited (by Invitation Only)

PRESENTER (FIRST NAME ONLY): Hanjong

PRESENTER (LAST NAME ONLY): Paik

PRESENTER (INSTITUTION ONLY): University of Oklahoma

PRESENTER (EMAIL ONLY): paik@ou.edu

PRESENTER (COUNTRY ONLY): United States

DATE/TIME SUBMITTED: September 24, 2023, 07:43 PM

ABSTRACT BODY:

Abstract Body: We focus on exploring the intricate phase diagram of the Sr-Ge-O thin film system using molecular-beam epitaxy. Under ultra-high vacuum (UHV) conditions, we employed a suboxide GeO-flux and elemental Sr-flux to successfully grow single-crystalline complex strontium germanates on $LaAlO_3$ (001) substrates. The growth of these films in a single phase is highly dependent on the ratio of Sr-flux/GeO-flux and substrate temperature. Through in-situ RHEED observation, we observed a transition in growth mode from an initial complicated 3D nucleation-dominant growth to a smoother 2D layer growth, which was consistent with Atomic Force Microscopy (AFM) observations. XRD measurements confirmed the presence of single-phase Sr_2GeO_4-like thin films with an out-of-plane d-space of 4.05 Å, indicating a strain-relaxed growth mechanism as evidenced by wider ω-rocking curves. Large area reciprocal space mapping revealed a unique domain structure. STEM imaging further demonstrated the epitaxial relationship between the thin films and $LaAlO_3$ (001) substrate, exhibiting a relatively sharp interface structure. This complex germanate system holds promise as a potential candidate for high-mobility oxide semiconductors in power electronics applications. We are currently conducting further studies involving optical and electrical doping to expand our understanding of this system's properties.

CONTROL ID: 3985247
CURRENT SYMPOSIUM: S3 - Advanced Electronic Materials: Processing and Structures and Properties and Applications
CURRENT SESSION: Advanced electronic materials, including ferroelectric, piezoelectric, dielectric, electrostrictive, electrocaloric and pyroelectric materials
TITLE: Effect of Co-Doping in $BaTiO_3$ Ceramics on Dielectric Properties
PRESENTATION TYPE: Poster
PRESENTER (FIRST NAME ONLY): Pooja
PRESENTER (LAST NAME ONLY): Kumari
PRESENTER (INSTITUTION ONLY): Oregon State Univ
PRESENTER (EMAIL ONLY): kumarip@oregonstate.edu
PRESENTER (COUNTRY ONLY): United States
DATE/TIME SUBMITTED: September 24, 2023, 07:45 PM
ABSTRACT BODY:
Abstract Body: Barium titanate (BT) with the perovskite type structure is a ferroelectric compound, which is widely used in dielectric-based applications, memory, sensors, and energy storage devices. Co-Doping in BT has dramatic influence on its electrical properties owing to the crossover from ferroelectric-to-ferroelectric relaxor. This study is focused on the effects of co-doping BT with rare earth metals with the goal of preparing a dielectric material with a diffuse phase transition to maintain permittivity over a broad temperature range. Standard solid-state synthesis was used to prepare ceramics samples which were characterized using X-ray diffraction, SEM, density, and dielectric measurements. The results of this study will help guide future work in the development of new high-permittivity dielectrics for MLCC applications.

CONTROL ID: 3985254
CURRENT SYMPOSIUM: S14 - Emerging Semiconductor Materials and Interfaces
CURRENT SESSION: WBG and UWBG materials - synthesis, characterization, theory
TITLE: New Nitride Semiconductors and Heterostructures
PRESENTATION TYPE: Invited (by Invitation Only)
PRESENTER (FIRST NAME ONLY): Joseph
PRESENTER (LAST NAME ONLY): Casamento
PRESENTER (INSTITUTION ONLY): Massachusetts Institute of Technology
PRESENTER (EMAIL ONLY): jac93@mit.edu
PRESENTER (COUNTRY ONLY): United States
DATE/TIME SUBMITTED: September 25, 2023, 09:29 PM
ABSTRACT BODY:
Abstract Body: Nitride semiconductors find significant usage in light emitting diodes and lasers, high power devices, radiofrequency power amplifiers, and bulk and surface acoustic wave resonators. To extend these technologies to higher voltage, frequency, and wavelength regimes, enhanced performance in the form of new materials and heterostructures is needed. In addition, the recent emergence of ferroelectric behavior draws nitrides into memory applications. This talk discusses growth, characterization, and device fabrication of two newer ternary nitride alloys: aluminum scandium nitride and aluminum boron nitride. Specifically, the aspects of epitaxial growth and lattice-matching with gallium nitride, their incoproration in nitride high electron mobility transistors (HEMTs) and the new functionalities and device architectures that stem from their high-K dielectric and ferroelectric behavior are discussed. Aluminum scandium nitride based ferroelectric HEMTs with operating frequencies in excess of 100 GHz and ultra-thin aluminum boron nitride ferroelectric capacitors with operating voltages of less than 10V is highlighted. Additionally, the challenges and opportunities of generating heterostructures with different deposition techniques, such as sputtering and molecular beam epitaxy (MBE), is discussed. A comparison to existing ternary nitride semiconductors such as indium aluminum nitride and alluminum gallium nitride will also be presented.

CONTROL ID: 3985255
CURRENT SYMPOSIUM: S14 - Emerging Semiconductor Materials and Interfaces
CURRENT SESSION: 2D materials - synthesis, characterization, theory
TITLE: Bilayer Graphene Grown in Frank van der Merwe Mode and Its Machine-Learning-Assisted Characterization
PRESENTATION TYPE: Invited (by Invitation Only)
PRESENTER (FIRST NAME ONLY): Haozhe
PRESENTER (LAST NAME ONLY): Wang
PRESENTER (INSTITUTION ONLY): Duke University
PRESENTER (EMAIL ONLY): haozhe.wang@duke.edu
PRESENTER (COUNTRY ONLY): United States
DATE/TIME SUBMITTED: September 24, 2023, 10:53 PM
ABSTRACT BODY:
Abstract Body: Bilayer graphene has become a rising star in the discovery of unconventional physics. While the number of exciting physical phenomena observed in bilayer graphene increases, a significant gap persists in transforming these discoveries into useful applications, owing to the small-scale samples obtained via the top-down approach. We realized a layer-by-layer (that is, Frank van der Merwe) growth mode in large-scale bilayer graphene, with no island impurities, which is unprecedented in any van der Waals-stacked materials. Owing to the counter-intuitive growth principle in chemical vapor deposition graphene, we proposed a new physical quantity, named "interface adhesive energy," that can be used to predict the growth mode. We show, through first-principles calculations, that this new physical quantity is tunable. We have thus realized the classical Frank van der Merwe growth mode in graphene. The characterization of graphene is a historical problem. Since the first report of graphene, researchers have been showing a few characterization images and spectra of a material, based on random sampling. The situation became ambiguous for the large-scale sample when massive data are involved. We have invented a machine-learning-assisted tool for Raman analysis for the precise characterization of stacking order and layer number of our graphene grown in Frank van der Merwe mode.

CONTROL ID: 3985276

CURRENT SYMPOSIUM: S8 - Data-driven and model-supported structure-property relationships in complex electroceramics

CURRENT SESSION: Model-supported advanced nano- and microscale characterization methods

TITLE: Integrating multiscale microscopy and machine learning to understand low-temperature degradation in yttria-stabilized zirconia

PRESENTATION TYPE: Invited (by Invitation Only)

PRESENTER (FIRST NAME ONLY): Baishakhi

PRESENTER (LAST NAME ONLY): Mazumder

PRESENTER (INSTITUTION ONLY): University at Buffalo - The State University of New York, University at Buffalo - The State University of New York, Buffalo, NY, US, academic

PRESENTER (EMAIL ONLY): baishakh@buffalo.edu

PRESENTER (COUNTRY ONLY): United States

DATE/TIME SUBMITTED: September 24, 2023, 09:46 PM

ABSTRACT BODY:

Abstract Body: Zirconia, a ceramic steel renowned for its remarkable properties, faces a formidable challenge in maintaining its structural integrity. The root cause of this challenge lies in the intricate management of atomistic characteristics, encompassing impurity control and vacancy concentrations, culminating in a phenomenon known as low-temperature degradation (LTD), primarily attributed to the tetragonal to monoclinic phase transition.

This presentation embarks on a journey to unravel the profound mechanisms underlying LTD in yttria-stabilized zirconia. To accomplish this ambitious task, we employ a cutting-edge arsenal of advanced microscopy techniques, including Atom Probe Tomography and Transmission Electron Microscopy. Furthermore, we harness the power of machine learning to dissect and model the complex data extracted from these microscopic observations. Our multidisciplinary, multiscale approach offers a comprehensive understanding of LTD, shedding light on its intricate dynamics and pivotal factors. Through our research, we reveal invaluable insights that are indispensable for the development of zirconia-based materials with enhanced durability. These discoveries hold the potential to revolutionize their practical applications across diverse scientific and industrial domains.

CONTROL ID: 3985288

CURRENT SYMPOSIUM: S13 - Frontiers in Ferroic oxides: Synthesis, Properties, and Applications

CURRENT SESSION: Electrical control of magnetism and spin textures

TITLE: Elastically tunable emergent magnetic state in Sr_2IrO_4 by strain

PRESENTATION TYPE: Contributed (Oral)

PRESENTER (FIRST NAME ONLY): Shashi

PRESENTER (LAST NAME ONLY): Pandey

PRESENTER (INSTITUTION ONLY): University of Tennessee

PRESENTER (EMAIL ONLY): shashi.sapiens@gmail.com

PRESENTER (COUNTRY ONLY): United States

DATE/TIME SUBMITTED: September 24, 2023, 11:07 PM

ABSTRACT BODY:

Abstract Body: Iridates are one of the highly investigated transitional metal oxides because of the unique combination of the electron-electron and strong spin-orbit interaction. A quasi-two-dimensional J_{eff} =1/2 canted antiferromagnetic (AF) Mott insulator Sr_2IrO_4 is a prominent example, which have a layered structure very similar to the parent phase of weakly spin-orbit-coupled high-Tc cuprates. However, due to the built-in spin-orbit entanglement, the J_{eff} = 1/2 moments can form significant inter-site quadrupoles in contrast to the S=1/2 moments of Cu ions. The resulting magnetoelastic coupling leads to spontaneous tetragonal symmetry breaking by the AF order in the B_{1g} channel. In the experiment, we compared the elasto-responses of the AF order to in-situ B_{1g} and B_{2g} strains that have orthogonally symmetric configurations. While the B1g strain efficiently detwins the spontaneous AF domains, with the B2g strain, a new states emerges that break the translational symmetry along the c-axis [1]. Our model analysis shows that such an emergent state is driven by an unusual quartic interaction of B_{2g} symmetry, competing with the intrinsic B_{1g} anisotropy, and can be elastically tuned by the applied strain.

1. S. Pandey et. al., Controllable Emergent Spatial Spin Modulation in Sr2IrO4 by In Situ Shear Strain, Phys. Rev. Lett. 129, 027203 (2022).

CONTROL ID: 3985291
CURRENT SYMPOSIUM: S1: Characterization of Structure-Property Relationships in Functional Ceramics
CURRENT SESSION: Integrating machine learning into the structural measurement workflow
TITLE: High Performance Computing and AI Enabled Materials Characterization and Experimental Automation
PRESENTATION TYPE: Invited (by Invitation Only)
PRESENTER (FIRST NAME ONLY): Mathew
PRESENTER (LAST NAME ONLY): Cherukara
PRESENTER (INSTITUTION ONLY): Argonne National Laboratory
PRESENTER (EMAIL ONLY): mcherukara@anl.gov
PRESENTER (COUNTRY ONLY): United States
DATE/TIME SUBMITTED: September 24, 2023, 10:25 PM
ABSTRACT BODY:
Abstract Body: The capabilities provided by next generation light sources along with the development of new characterization techniques and detector advances are revolutionizing materials characterization (metrology) by providing the ability to perform scale-bridging, multi-modal materials characterization under in-situ and operando conditions. For example, providing the ability to image in 3D large fields of view (~mm^3) at high resolution (<10 nm), while simultaneously acquiring information about structure, strain, elemental composition, oxidation state, photovoltaic response etc.

However, these novel capabilities dramatically increase the complexity and volume of data generated. Conventional data processing and analysis methods become infeasible in the face of such large and varied data streams. The use of AI/ML methods is becoming indispensable for real-time analysis, data abstraction and decision making at advanced, high-data rate instruments. I will describe how high-performance computing (HPC) along with AI on edge devices enables real-time data analysis and self-driving experiments, creating the next generation of AI-powered materials characterization tools.

References:
Kandel, Saugat, et al., Nature Communications, 14, 5501 (2023).
Babu, Anakha V., et al. arXiv:2209.09408, Nature Communications, in press (2023).
Cherukara, Mathew J., et al. Applied Physics Letters 117.4 (2020): 044103.

CONTROL ID: 3985311

CURRENT SYMPOSIUM: S4 - Complex Oxide Thin Films and Heterostructures: From Synthesis to Strain/Interface-engineered Emergent Properties

CURRENT SESSION: Strain, microstructures and functionality tuning in superlattices, vertical nanocomposites, and high entropy complex oxides

TITLE: Recent Progress and Future Directions for Oxide-based and Nitride-based Vertically Aligned Nanocomposites

PRESENTATION TYPE: Invited (by Invitation Only)

PRESENTER (FIRST NAME ONLY): Haiyan

PRESENTER (LAST NAME ONLY): Wang

PRESENTER (INSTITUTION ONLY): Purdue University

PRESENTER (EMAIL ONLY): hwang00@purdue.edu

PRESENTER (COUNTRY ONLY): United States

DATE/TIME SUBMITTED: September 24, 2023, 11:13 PM

ABSTRACT BODY:

Abstract Body: Two-phase nanocomposites have gained significant research interests because of their multifunctionalities, tunable geometries and potential device applications. Different from the prior demonstrated oxide-oxide 2-phase vertically aligned nanocomposites (VANs), coupling either oxides or nitrides with metals presents high potentials of building alternative hybrid plasmonic metamaterials towards chemical sensing, tunable plasmonics, and nonlinear optics. Unique advantages, including distinct atomic interface, excellent crystalline quality, large-scale surface coverage and durable solid-state platform, address the high demands in new hybrid metamaterial designs for versatile optical material needs. This talk discusses some of the recent progress in oxide-metal, nitride-metal, nitride-oxide and some of the complex three phase VANs, with unique bottom-up self-assembly. Unique properties including plasmonic resonance tuning, optical anisotropy, magnetic anisotropy, and strong magneto-optical coupling, are demonstrated in these systems. Future research directions are also discussed with the goals to achieve highly ordered secondary phase and integration on device-relevant substrates.

The work of oxide-metal VAN is funded by the DOE BES program DE-SC0020077. The work of nitride-based VAN is funded by the U.S. National Science Foundation, DMR-2016453.

CONTROL ID: 3985347
CURRENT SYMPOSIUM: S1: Characterization of Structure-Property Relationships in Functional Ceramics
CURRENT SESSION: Advances in scattering, imaging, and analytical techniques
TITLE: Optimization of SiC/CNFs Composites for High-Temperature Applications: Experimental and Molecular Dynamics Study
PRESENTATION TYPE: Contributed (Oral)
PRESENTER (FIRST NAME ONLY): Mubina
PRESENTER (LAST NAME ONLY): Shaik
PRESENTER (INSTITUTION ONLY): National Institute of Technology Andhra Pradesh
PRESENTER (EMAIL ONLY): mubinashaik93@gmail.com
PRESENTER (COUNTRY ONLY): India
DATE/TIME SUBMITTED: September 25, 2023, 07:14 AM
ABSTRACT BODY:
Abstract Body: Silicon carbide-based composites have found applications in various fields due to their unique thermal and mechanical properties. Carbon nanofibers (CNFs) are known for their exceptional thermal, mechanical, and electrical properties, making them excellent potential additives and reinforcements. CNFs-reinforced SiC composites exhibit exceptional high-temperature properties, making them suitable materials for nuclear applications.

In this study, a Gray Relational Analysis (GRA)-based Taguchi statistical analysis was employed to optimize the processing parameters of CNF-reinforced SiC composites. An advanced spray freeze granulation technique was adopted for the consolidation of SiC/CNFs powders, followed by Pressurless sintering at 2150°C/1hr under Ar atmosphere. The physical, thermal, and mechanical properties of the samples were measured and analyzed using GRA-based Taguchi analysis.

Molecular Dynamics (MD) simulation is an important method for studying the atomic-scale properties of materials. In the context of cladding materials, thermal conductivity is a crucial physical property. Therefore, in this study, the thermal properties of SiC/CNF composites were computed up to 1000°C using the conventional Laser Flash technique and molecular dynamic simulations (MD) via the Green–Kubo or non-equilibrium MD (NEMD) methods. These results were then compared.

CONTROL ID: 3985396

CURRENT SYMPOSIUM: S1: Characterization of Structure-Property Relationships in Functional Ceramics

CURRENT SESSION: Addressing open questions in functional ceramics

TITLE: Structure-property relationships in polar perovskite oxides

PRESENTATION TYPE: Contributed (Oral)

PRESENTER (FIRST NAME ONLY): Neamul

PRESENTER (LAST NAME ONLY): Khansur

PRESENTER (INSTITUTION ONLY): Friedrich-Alexander Universität Erlangen-Nürnberg

PRESENTER (EMAIL ONLY): neamul.khansur@fau.de

PRESENTER (COUNTRY ONLY): Germany

DATE/TIME SUBMITTED: September 25, 2023, 04:55 AM

ABSTRACT BODY:

Abstract Body: Energy generation, distribution, and storage remain the key elements to decrease CO_2 emissions to ~45% by 2030 and reach net zero by 2050. Therefore, it is crucial to develop effective technologies for simultaneous conversion of energy from multiple sources as well as energy storage. Perovskite oxides are one of the most versatile material classes for multifunctional energy conversion systems. This includes the potential to convert mechanical vibrations into electrical energy through the direct piezoelectric effect, the conversion of thermal fluctuations into electrical energy via the pyroelectric effect, and the possibility of using the internal electric fields present in polar oxide materials to influence electron-hole recombination in solar-cell devices or chemical reactions, such as water-splitting. The functional properties are governed by their underlying structure. Importantly, however, the structure-property relationships are also influenced by the processing conditions, defect concentrations, post-processing thermal treatment, as well as external stimuli such as thermal, electrical, and/or mechanical fields. As such, understanding the structural contributions to the macroscopic functional properties is critical to tune the multifunctionality. In this contribution, structure-property relationships of polar perovskite oxides under external applied electric and mechanical fields will be discussed.

CONTROL ID: 3985417
CURRENT SYMPOSIUM: S5 - Controlling Grain Boundary Structure, Chemistry, and their Network as a Function of Mtl Processing
CURRENT SESSION: Defect and segregation behavior
TITLE: Intergranular Chemistry and Functionality of Segregates inside Metal Oxide Nanoparticle Derived Ceramics
PRESENTATION TYPE: Contributed (Oral)
PRESENTER (FIRST NAME ONLY): Oliver
PRESENTER (LAST NAME ONLY): Diwald
PRESENTER (INSTITUTION ONLY): Paris-Lodron Universität Salzburg
PRESENTER (EMAIL ONLY): oliver.diwald@plus.ac.at
PRESENTER (COUNTRY ONLY): Austria
DATE/TIME SUBMITTED: September 25, 2023, 07:28 AM
ABSTRACT BODY:
Abstract Body: This study targets intergranular domains in ceramic MgO based microstructures with engineered optical, magnetic or semiconducting properties. For this purpose, we used MgO-based nanocomposite powders, which were grown by flame spray pyrolysis and processed by powder compaction and application of dedicated annealing cycles. During synthesis included Ba^{2+}, In^{3+} and Fe^{3+} admixtures served as precursors for the chemical engineering of the intergranular phase.
Annealing-induced ion exsolution from the nanocrystal volume into the grain boundaries and intergranular regions were found to also affect grain coarsening and densification behaviour of the powders, effects that were compared for the first time using an integrated characterization approach. A comprehensive structural analysis using electron microscopy and X-ray diffraction allowed us to draw conclusions about the structure-property relationships controlled by impurity dispersion within the network of MgO grains. The functionalization of the intergranular region with BaO (for interface-sensitive luminescence), with $MgFe_2O_4$ (for magnetism), or with $MgIn_2O_4$ (for electronic conductivity) will be demonstrated by relevant physical measurements. Nature and reactivity of the functional admixture and its respective loading inside the structures will be discussed.

CONTROL ID: 3985456
CURRENT SYMPOSIUM: S7 - Defects and Transport in Ceramics
CURRENT SESSION: Point defect segregation to or depletion from dislocations, surfaces, grain boundaries, and interfaces
TITLE: Spontaneous Polarization Versus Point Defect Chemistry in $BaTiO_3$ Nanocrystals
PRESENTATION TYPE: Invited (by Invitation Only)
PRESENTER (FIRST NAME ONLY): Oliver
PRESENTER (LAST NAME ONLY): Diwald
PRESENTER (INSTITUTION ONLY): Paris-Lodron Universität Salzburg
PRESENTER (EMAIL ONLY): oliver.diwald@plus.ac.at
PRESENTER (COUNTRY ONLY): Austria
DATE/TIME SUBMITTED: September 25, 2023, 09:27 AM
ABSTRACT BODY:
Abstract Body: Knowledge about the emergence and depletion of point defects in $BaTiO_3$ (BTO) nanostructures during materials processing is key to our understanding of their later activity as components in functional dielectric devices or as photocatalysts. We analyzed the influence of powder annealing on spontaneous polarization and the potential formation of recombination centers in $BaTiO_3$ nanocrystals. For this purpose, the contribution of i) tetragonal distortion, responsible for spontaneous polarization, and ii) point defects on charge separation and recombination within the $BaTiO_3$ nanocrystals was investigated using an integrated characterization approach. Electron paramagnetic resonance in combination with O_2 photoadsorption experiments show for the first time the beneficial effect of the BTO perovskite nanocrystal lattice on charge separation. On the other hand, charge separation is considerably hindered within BTO nanoparticles annealed to temperatures ≥ 600 °C, due to the formation of Ba–O divacancies that act as recombination centers. The opposing interplay between tetragonal distortion and annealing-induced defect formation inside the lattice highlights the importance of defect engineering within perovskite nanoparticles.

CONTROL ID: 3985486
CURRENT SYMPOSIUM: S10 - Fundamental mechanisms in materials for next generation computing devices
CURRENT SESSION: Phase transitions in 2D materials and devices
TITLE: Novel mechanisms for ferroelectricity and antiferroelectricity near the atomically thin limit
PRESENTATION TYPE: Invited (by Invitation Only)
PRESENTER (FIRST NAME ONLY): Salvador
PRESENTER (LAST NAME ONLY): Barraza-Lopez
PRESENTER (INSTITUTION ONLY): University of Arkansas
PRESENTER (EMAIL ONLY): sbarraza@uark.edu
PRESENTER (COUNTRY ONLY): United States
DATE/TIME SUBMITTED: September 25, 2023, 09:29 AM
ABSTRACT BODY:
Abstract Body: New material platforms for ferroelectricity and antiferroelectricity invite to rethink of the physical mechanisms that underpin their behavior. In this talk, I will review some novel 2D ferroelectric and antiferroelectric materials, and will contrast their behavior with that observed in oxide ferroelectrics. The novel behaviors include a strong coupling to distortions of the lattice, the possibility of achieving ferroelectricity in binary or elemental compounds, and antiferroelectric behaviors linked to topological atomic phases.

CONTROL ID: 3985541
CURRENT SYMPOSIUM: S13 - Frontiers in Ferroic oxides: Synthesis, Properties, and Applications
CURRENT SESSION: Low dimension ferroic oxides
TITLE: Order Parameters in the Infinite Layer and Reduced Nickelates NdNiO2+x
PRESENTATION TYPE: Invited (by Invitation Only)
PRESENTER (FIRST NAME ONLY): Kyle
PRESENTER (LAST NAME ONLY): Shen
PRESENTER (INSTITUTION ONLY): Cornell University
PRESENTER (EMAIL ONLY): kmshen@cornell.edu
PRESENTER (COUNTRY ONLY): United States
DATE/TIME SUBMITTED: September 25, 2023, 10:07 AM
ABSTRACT BODY:
Abstract Body: The infinite layer nickelates present an opportunity to compare and contrast against their cuprate cousins, when determining the necessary ingredients for unconventional superconductivity and the role of competing order parameters such as charge order and superconductivity. Resonant x-ray scattering experiments have reported charge ordering with three unit cell periodicity, suggesting that an interplay exists between charge order and superconductivity much like the cuprates. However, there have also been conflicting reports about the existence of charge order in differently prepared samples, suggesting that charge order could be stabilized or enhanced by the presence of disorder, or some unknown variable. To investigate this, we synthesized a sequence of $NdNiO_2$ samples with varying levels of disorder by systematically varying the reduction conditions of different samples. Resonant x-ray scattering revealed that the strength of the charge order peak depended strongly on reducing conditions and sample inhomogeneity. We also observe that in samples with the lowest residual resistivity, traces of superconductivity emerge, even without hole doping. We discuss the implications on the nickelate phase diagram and high-temperature superconductivity in general.

CONTROL ID: 3985549
CURRENT SYMPOSIUM: S3 - Advanced Electronic Materials: Processing and Structures and Properties and Applications
CURRENT SESSION: Advanced electronic materials, including ferroelectric, piezoelectric, dielectric, electrostrictive, electrocaloric and pyroelectric materials
TITLE: The Effect of Ca, Sr, and Ba-doping on the Dielectric and Polarization Properties of PZT 95/5 Ceramics
PRESENTATION TYPE: Contributed (Oral)
PRESENTER (FIRST NAME ONLY): Eric
PRESENTER (LAST NAME ONLY): Neuman
PRESENTER (INSTITUTION ONLY): Sandia National Laboratories
PRESENTER (EMAIL ONLY): ewneuma@sandia.gov
PRESENTER (COUNTRY ONLY): United States
DATE/TIME SUBMITTED: September 25, 2023, 10:28 AM
ABSTRACT BODY:
Abstract Body: Dense PZT 95/5-2Nb was prepared by the solid-state synthesis route using traditional ceramic processing methods. Ca, Sr, and Ba were systematically added in amounts up to 0.01 mol. % to the PZT. The sintered material was characterized using ICP-OES, XRD, and SEM/EDS, impedance spectroscopy, P-E-ε hysteresis looping, and pressure induced depolarization. All ceramics were ~98% dense with an average grain size of 6 μm. Curie temperature decreased with all dopant additions, while the rhombohedral FE_{low} to FE_{high} phase transition increased with Ca doping and decreased with Sr and Ba doping. Polarization was increased with Sr doping and decreased with Ca and Ba doping, while coercive field decreased for all dopants. The effect of composition on the processing, properties, and FE-AFE phase transition, and tolerance factor will be discussed.

Sandia National Laboratories is a multimission laboratory managed and operated by National Technology & Engineering Solutions of Sandia, LLC, a wholly owned subsidiary of Honeywell International Inc., for the U.S. Department of Energy's National Nuclear Security Administration under contract DE-NA0003525.

CONTROL ID: 3905507
CURRENT SYMPOSIUM: S3 - Advanced Electronic Materials: Processing and Structures and Properties and Applications
CURRENT SESSION: Advanced electronic materials, including ferroelectric, piezoelectric, dielectric, electrostrictive, electrocaloric and pyroelectric materials
TITLE: Electronic and Microstructure effects of Aluminum Additions in Niobium-Doped Tin-Modified PZT 95/5
PRESENTATION TYPE: Contributed (Oral)
PRESENTER (FIRST NAME ONLY): Patrick
PRESENTER (LAST NAME ONLY): Price
PRESENTER (INSTITUTION ONLY): Sandia National Laboratories
PRESENTER (EMAIL ONLY): pmpric@sandia.gov
PRESENTER (COUNTRY ONLY): United States
DATE/TIME SUBMITTED: September 25, 2023, 10:39 AM
ABSTRACT BODY:
Abstract Body: The effect of aluminum addition in Nb-doped, Sn-modified PZT 95/5 (PSZT(Nb)) on microstructure, composition, dielectric, and ferroelectric properties are explored. Compositions of PSZT(Nb) were batched with three levels of aluminum content, 0, 1000, and 3000 ppm. This work builds on the previously reported sintering dynamics in this system and discusses the trends observed between aluminum content, microstructure development, electronic properties, and composition. Electron microscopy and elemental analysis suggests that added aluminum does not incorporate into the solid solution, instead migrating to grain boundaries and concentrating at triple points. This supports the observation of varied sintering kinetics with increasing aluminum content and leads to various trends in electrical properties.

Sandia National Laboratories is a multimission laboratory managed and operated by National Technology & Engineering Solutions of Sandia, LLC, a wholly owned subsidiary of Honeywell International Inc., for the U.S. Department of Energy's National Nuclear Security Administration under contract DE-NA0003525.

CONTROL ID: 3985572

CURRENT SYMPOSIUM: S3 - Advanced Electronic Materials: Processing and Structures and Properties and Applications

CURRENT SESSION: Advanced electronic materials, including ferroelectric, piezoelectric, dielectric, electrostrictive, electrocaloric and pyroelectric materials

TITLE: Exploration of magnetron sputtered κ-Ga_2O_3 for ferroelectricity

PRESENTATION TYPE: Contributed (Oral)

PRESENTER (FIRST NAME ONLY): Emilee

PRESENTER (LAST NAME ONLY): Fortier

PRESENTER (INSTITUTION ONLY): The Pennsylvania State University

PRESENTER (EMAIL ONLY): Ecfortie@gmail.com

PRESENTER (COUNTRY ONLY): United States

DATE/TIME SUBMITTED: September 25, 2023, 09:36 PM

ABSTRACT BODY:

Abstract Body: Metastable κ-Ga_2O_3 is an orthorhombic ordered substructure of the hexagonal ε phase of Ga_2O_3. Recently, κ-Ga_2O_3 has become the subject of increasing research interest due to its spontaneous polarization, strong piezoelectricity, and reported ferroelectric properties. Further support for potential of κ-Ga_2O_3 as a novel ferroelectric is provided by research demonstrating the switching of isomorphic $Pna2_1$ Fe_2O_3 and $Ga_xFe_{2-x}O_3$. Currently, no reports have demonstrated phase-pure synthesis of epsilon or kappa phase Ga_2O_3 by magnetron sputtering, which is an industrially relevant method of film preparation.

Two bulk ceramic sputter targets of compositions $(Ga_{1-x}M_x)_2O_3$ (M=Sn,Sc) have been prepared by shaker-milling and sintering methods and have used as source materials for exploring the material systems Sn and Sc incorporated Ga_2O_3 by RF magnetron sputtering. The role of Sn and Sc incorporation on the stabilization of thin film κ-Ga_2O_3 has been investigated and the growth space of κ-Ga_2O_3 on metallic bottom electrodes by RF magnetron sputtering has been mapped. This talk will report on the best practices for the preparation of bulk $(Ga_{1-x}M_x)_2O_3$ (M=Sn,Sc) as well as the rf magnetron sputtering conditions that yield best phase purity and electrical properties relevant to the ferroelectric capabilities of κ-Ga_2O_3.

CONTROL ID: 3985582
CURRENT SYMPOSIUM: S1: Characterization of Structure-Property Relationships in Functional Ceramics
CURRENT SESSION: Integrating machine learning into the structural measurement workflow
TITLE: Semi and self supervised approaches to space group and bravais lattice determination
PRESENTATION TYPE: Invited (by Invitation Only)
PRESENTER (FIRST NAME ONLY): William
PRESENTER (LAST NAME ONLY): Ratcliff
PRESENTER (INSTITUTION ONLY): NIST
PRESENTER (EMAIL ONLY): william.ratcliff@nist.gov
PRESENTER (COUNTRY ONLY): United States
DATE/TIME SUBMITTED: September 25, 2023, 10:55 AM
ABSTRACT BODY:
Abstract Body:
During this talk, I will discuss our work [1] to use neural networks to automatically classify Bravais lattices and space-groups from neutron powder diffraction data. Our work classifies 14 Bravais lattices and 144 space groups. The novelty of our approach is to use semi-supervised and self-supervised learning to allow for training on data sets with unlabelled data as is common at user facilities. We achieve state of the art results with a semi-supervised approach. Our accuracy for our self-supervised training is comparable to that with a supervised approach.

Support for Satvik Lolla was provided by the Center for High Resolution Neutron Scattering, a partnership between the National Institute of Standards and Technology and the National Science Foundation under Agreement No. DMR-2010792.

[1] Satvik Lolla Et al, Journal of Applied Crystallography 55 (2022) https://doi.org/10.1107/S1600576722006069

CONTROL ID: 3985616
CURRENT SYMPOSIUM: S7 - Defects and Transport in Ceramics
CURRENT SESSION: Defect-mediated properties
TITLE: Understanding Phenomenological Properties of Dielectrics and Thermoelectrics in Terms of Defect Chemistry
PRESENTATION TYPE: Invited (by Invitation Only)
PRESENTER (FIRST NAME ONLY): Soonil
PRESENTER (LAST NAME ONLY): Lee
PRESENTER (INSTITUTION ONLY): Changwon National University
PRESENTER (EMAIL ONLY): leesoonil@changwon.ac.kr
PRESENTER (COUNTRY ONLY): Korea (the Republic of)
DATE/TIME SUBMITTED: September 25, 2023, 12:39 PM
ABSTRACT BODY:
Abstract Body: Many of the advanced properties observed in the materials we use are extrinsic and can be attributed to the presence of defects, such as dopants, solid-solutions, domains, inclusions, grain boundaries, localized phases, pores, and so on. Point defects, especially in electroceramics and semiconductors, lead to significantly different electrical, thermal, magnetic, and optical properties. Certain defects, like anion and cation vacancies, can be problematic for dielectric applications but prove beneficial for thermoelectrics, electrolytes, and other applications. The contribution of these defects has been determined through various measurements. Point defects in ABO_3–based ceramics were investigated through high temperature equilibrium conductivity (HiTEC) as a function of oxygen partial pressure. Through engineering defects, various phenomina in dielectric, piezoelectric, and thermoelectric properties were confirmed and understood in terms of defect chemistry.

CONTROL ID: 3985620
CURRENT SYMPOSIUM: S11 - In situ/operando Characterization of Nanomaterials
CURRENT SESSION: Study of nanomaterial-based catalysts at reactive environments
TITLE: Operando neutron total scattering studies of compositionally complex defect fluorite nanocatalysts
PRESENTATION TYPE: Invited (by Invitation Only)
PRESENTER (FIRST NAME ONLY): Katharine
PRESENTER (LAST NAME ONLY): Page
PRESENTER (INSTITUTION ONLY): University of Tennessee
PRESENTER (EMAIL ONLY): kpage10@utk.edu
PRESENTER (COUNTRY ONLY): United States
DATE/TIME SUBMITTED: September 25, 2023, 12:14 PM
ABSTRACT BODY:
Abstract Body: Integration of renewable feedstocks into the current energy infrastructure drives the development of heterogenous catalysts and electrocatalysts that can maintain high surface areas and selective catalytic/electrocatalytic activities under a variety of challenging chemical and environmental conditions. Recently, several compositionally complex oxides have been reported with intriguing and tunable catalytic properties. Structural probes capable of capturing chemically reactive catalyst surfaces (including correlated defects and chemical short-range order) in real time and at elevated temperatures, remain extremely limited. Neutrons offer advantages of light atom sensitivity and superior penetration of sample environments for operando studies, while total scattering methods provide information on local atomic structure motifs at nanoparticle surfaces and interiors. We present recent studies on compositionally complex Rare Earth cerate and cerate zirconate catalysts and electrocatalysts, highlighting the importance of studying structure-property characteristics under real-world reaction conditions. Emerging opportunities for operando investigations with a hazardous gas handling system currently under construction at the Spallation Neutron Source at Oak Ridge National Laboratory will be presented.

CONTROL ID: 3985646
CURRENT SYMPOSIUM: S16 - New Frontiers and Science in Additive Manufacturing of Ceramic Materials
CURRENT SESSION: New approaches for AM processes of ceramics and ceramic-matrix composites
TITLE: High-Speed and High-resolution 3D Printing of Neodymium-doped Yttrium Aluminum Garnet Ceramics with Optical Gain Properties
PRESENTATION TYPE: Invited (by Invitation Only)
PRESENTER (FIRST NAME ONLY): Xiangfan
PRESENTER (LAST NAME ONLY): Chen
PRESENTER (INSTITUTION ONLY): Arizona State University
PRESENTER (EMAIL ONLY): Xiangfan.Chen@asu.edu
PRESENTER (COUNTRY ONLY): United States
DATE/TIME SUBMITTED: September 25, 2023, 12:00 PM
ABSTRACT BODY:
Abstract Body: Polycrystalline transparent ceramics doped with rare-earth ions are wildly used as lasing media. Specifically, neodymium-doped yttrium aluminum garnet (Nd: YAG), which can emit light with a wavelength of 1064 nm, has broad applications in medical, manufacturing, and automotive industry fields.However, conventional fabrication methods for Nd: YAG lasing media, such as powder pressing or gel casting followed by post-sintering, have a relatively long timescale. Besides, it is complicated to customize the 3D structures of products in these approaches. Herein, we develop a new method to fabricate multi-scaled Nd: YAG-based lasing media by using custom-made micro-continuous liquid interface printing followed by post-sintering. After optimization, the printing speed reaches 10 $\mu m*s^{-1}$ with a resolution of 5.8 $\mu m*pixel^{-1}$. The resultant samples doped with Nd element exhibited satisfactory photoluminescence (PL) performance, emitting light concentrated at 1064 nm upon stimulation by a 532 nm laser excitation. Moreover, PL imaging was captured to visualize the emitted light from the output end of 3D Nd: YAG ceramic structures when subjected to irradiation at the input end. Those comprehensive efforts collectively demonstrate the possibility of prototyping Nd: YAG ceramic gain media for diverse applications.

CONTROL ID: 3985649

CURRENT SYMPOSIUM: S4 - Complex Oxide Thin Films and Heterostructures: From Synthesis to Strain/Interface-engineered Emergent Properties

CURRENT SESSION: Strain- and interface-controlled device performance in energy harvesting, memories, sensors, etc.

TITLE: Impact of Electrodes on Retention Performance of Ferroelectric $Hf_{0.5}Zr_{0.5}O_2$ Capacitors

PRESENTATION TYPE: Contributed (Oral)

PRESENTER (FIRST NAME ONLY): Benjamin

PRESENTER (LAST NAME ONLY): Aronson

PRESENTER (INSTITUTION ONLY): University of Virginia

PRESENTER (EMAIL ONLY): ubz5mc@virginia.edu

PRESENTER (COUNTRY ONLY): United States

DATE/TIME SUBMITTED: September 25, 2023, 03:41 PM

ABSTRACT BODY:

Abstract Body: This work focuses on the polarization retention of 10 nm $Hf_{0.5}Zr_{0.5}O_2$ (HZO) thin film capacitors with TaN and tungsten electrodes. Polarization margin was quantitatively evaluated for same state and opposite state operations for up to 10^5 seconds at temperatures of 85, 105, and 150 °C. Both same state and opposite state polarization margins indicate no significant change for devices on tungsten electrodes up to 10^5 seconds up to 150 °C. In contrast, devices on tantalum nitrides showed a 50% polarization loss after 10^5 seconds at 150 °C. Imprint and thermal depolarization are two mechanisms of retention loss that were evaluated for devices on both electrodes. After every retention measurement polarization-electric field hysteresis and current loops were measured to quantify the degree of imprint in each sample. The devices with tantalum nitride capping electrodes underwent greater imprint due to a greater oxygen vacancy content, validated by photoluminescence measurements. Determining the polarization margin necessitates switching measurements which destroy the previous write. As such, pyroelectric coefficients were measured to determine if devices depolarized at elevated temperature bakes. While the tungsten devices did not depolarize, the devices bound by tantalum nitride electrode did undergo thermal depolarization, which accompanied the polarization imprint.

CONTROL ID: 3985658
CURRENT SYMPOSIUM: S5 - Controlling Grain Boundary Structure, Chemistry, and their Network as a Function of Mtl Processing
CURRENT SESSION: Nanocrystalline ceramics
TITLE: Nanostructural Effects and Superhardness in Tungsten Carbide
PRESENTATION TYPE: Invited (by Invitation Only)
PRESENTER (FIRST NAME ONLY): Kevin
PRESENTER (LAST NAME ONLY): Anderson
PRESENTER (INSTITUTION ONLY): US Naval Research Laboratory
PRESENTER (EMAIL ONLY): kpa951@gmail.com
PRESENTER (COUNTRY ONLY): United States
DATE/TIME SUBMITTED: September 25, 2023, 04:13 PM
ABSTRACT BODY:
Abstract Body: Industrial application of superhard materials (Vickers hardness, HV > 40 GPa) such as diamond and cubic boron nitride is limited by high costs and complex routes of synthesis. However, recent efforts have demonstrated the potential for alternative materials with similar hardness via microstructural refinement. Though broadly utilized for its hardness, tungsten carbide (WC) falls well short of qualification as a superhard material (HV ~ 26 GPa). In this work, a unique process known as environmentally controlled – pressure assisted sintering yields WC ceramics with HV as high as 39 GPa. The unprecedented WC hardness is attributed to the combined effects of record small grain size (26 nm) and nanoscale intragranular substructures resulting from lattice deformation. The demonstration of the superposition of multiple hardening mechanisms provides a potential generalized avenue to improve hardness of ceramics beyond grain size refinement, yielding new classes of superhard materials.

CONTROL ID: 3985660
CURRENT SYMPOSIUM: S12 - Magnetic and Superconducting Materials: From Basic Science to Applications
CURRENT SESSION: Other magnetic materials
TITLE: Efficient Control of 2D Magnets
PRESENTATION TYPE: Invited (by Invitation Only)
PRESENTER (FIRST NAME ONLY): Cheng
PRESENTER (LAST NAME ONLY): Gong
PRESENTER (INSTITUTION ONLY): University of Maryland
PRESENTER (EMAIL ONLY): gongc@umd.edu
PRESENTER (COUNTRY ONLY): United States
DATE/TIME SUBMITTED: September 25, 2023, 12:10 PM
ABSTRACT BODY:
Abstract Body: The emergent two-dimensional (2D) layered magnets provide ideal platforms to enable the atomically thin magneto-optical and magnetoelectric devices. Though many have envisioned that 2D magnets should allow efficient control of magnetism by a variety of external stimuli, true breakthroughs are still lacking, with limited proof-of-concept demonstrations reported thus far. There appear to be fundamental obstacles for efficient control, e.g., through electrical and optical means. In this talk I will analyze the challenges and present our theoretical and experimental progress on efficient electrical and optical control of 2D magnets. Specifically, the results show that the voltage of a few volts can effectively change the magnetic anisotropy of 2D magnets and the laser shinning of tens of uW/um^2 can effectively affect the domain behaviors of 2D magnets. These efficient controls of 2D magnets potentially open up new avenues towards low-power spintronics and photonics.

CONTROL ID: 3985668
CURRENT SYMPOSIUM: S2 - Advanced Microelectronics
CURRENT SESSION: Role of defects, interface and synthesis/fabrication process on device properties
TITLE: Synthesis and optoelectronic properties of wurtzite Al1-xGdxN heterostructural alloys
PRESENTATION TYPE: Contributed (Oral)
PRESENTER (FIRST NAME ONLY): Keisuke
PRESENTER (LAST NAME ONLY): Yazawa
PRESENTER (INSTITUTION ONLY): National Renewable Energy Laboratory
PRESENTER (EMAIL ONLY): Keisuke.Yazawa@nrel.gov
PRESENTER (COUNTRY ONLY): United States
DATE/TIME SUBMITTED: October 09, 2023, 02:33 PM
ABSTRACT BODY:
Abstract Body: AlN base materials have garnered attention in the field of microelectronics for their utility as semiconductors, piezoelectrics, and ferroelectrics. The introduction of rare earth elements into a ternary (Al,X)N matrix enables further functionalities such as optoelectronic behavior and magnetism. However, synthesizing a solid solution of the pure wurtzite phase in the alloy with a range of rare earth content poses a challenge because of a significant miscibility gap that can lead to spinodal decomposition.

We demonstrate synthesis of a single phase Al1-xGdxN heterostructural alloy with Gd content up to x = 0.24 using reactive sputtering. The high energy bombardment of the sputtered particles facilitates the formation of the metastable wurtzite phase. The bandgap of the wurtzite Al1-xGdxN decreases with Gd content. The deposited Al1-xGdxN films exhibit high resistivity, UV cathodoluminescence, and a photocurrent response under ultraviolet light illumination, indicating the material's potential for optoelectronic applications.

CONTROL ID: 3985669
CURRENT SYMPOSIUM: S18 - Structure and function in high-entropy, entropy-stabilized, and related oxides
CURRENT SESSION: Characterization and quantification of structure and properties
TITLE: Structure-Property Tuning in Compositionally Complex Pyrochlore and Spinel Oxide Families
PRESENTATION TYPE: Invited (by Invitation Only)
PRESENTER (FIRST NAME ONLY): Katharine
PRESENTER (LAST NAME ONLY): Page
PRESENTER (INSTITUTION ONLY): University of Tennessee
PRESENTER (EMAIL ONLY): kpage10@utk.edu
PRESENTER (COUNTRY ONLY): United States
DATE/TIME SUBMITTED: September 25, 2023, 12:30 PM
ABSTRACT BODY:
Abstract Body: Next generation materials of nearly every kind rely on chemical, electronic, and/or magnetic heterogeneity for creating, harnessing, and controlling functionality. High entropy oxides (HEOs), materials exhibiting a single-phase crystal structure containing five or more different cations on single crystallographic lattice sites, have attracted great interest in diverse fields because of their inherent opportunities to tailor and combine materials functionalities. The control of local order/disorder in the class is by extension a grand challenge towards realizing their vast potential. We present detailed investigation of the cation site preferences, chemical-short-range order, and electronic, thermal, and catalytic properties achieved in specific rare-earth and transition metal-based pyrochlore and spinel HEO families through compositional tuning and variation in synthesis/processing conditions. A combination of local to long-range electron, x-ray and neutron scattering probes and advanced theoretical modeling are employed to investigate their complex configurational diversity and associated structure-property trends. This work hints at the exquisite level of detail that may be required in computational and experimental approaches to guide structure-property tuning in emerging HEO materials. Current challenges and future opportunities in this arena will be discussed.

CONTROL ID: 3985709
CURRENT SYMPOSIUM: S14 - Emerging Semiconductor Materials and Interfaces
CURRENT SESSION: WBG and UWBG materials - synthesis, characterization, theory
TITLE: Growth and Properties of P-type α-Cr_2O_3 thin films for the High-temperature Operation of α-Cr_2O_3/β-Ga_2O_3 P-N Heterojunction Diodes
PRESENTATION TYPE: Contributed (Oral)
PRESENTER (FIRST NAME ONLY): Kingsley
PRESENTER (LAST NAME ONLY): Egbo
PRESENTER (INSTITUTION ONLY): National Renewable Energy Laboratory
PRESENTER (EMAIL ONLY): kingsley.egbo@nrel.gov
PRESENTER (COUNTRY ONLY): United States
DATE/TIME SUBMITTED: September 25, 2023, 04:00 PM
ABSTRACT BODY:
Abstract Body: Wide bandgap β-Ga2O3-based heterojunction diodes show significant promise for high-temperature applications due to their high barrier height and reduced leakage current. However, thermodynamic instability at the p-n heterojunction interface during high-temperature operation can lead to performance degradation. Here, we predict that the α-Cr_2O_3/β-Ga_2O_3 interface is thermodynamically stable, and then use this prediction to fabricate highly stable devices. Firstly, corundum structured chromium oxide (α -Cr_2O_3) was grown by pulsed laser deposition on different orientations of β-Ga_2O_3 and on reference c-Al_2O_3. While epitaxial α -Cr_2O_3 (006) is stabilized on c-Al_2O_3 and β-Ga_2O_3 (-201), growth on (001) was polycrystalline with negligible preferential orientation. Ti /Au (5 nm/100nm) was used to form stable Ohmic contacts to both α-Cr_2O_3 and β-Ga_2O_3. We then fabricate vertical heterojunction α -Cr_2O_3/β-Ga_2O_3 diodes on a 1μm thick β-Ga_2O_3 drift layer. A turn-on voltage, rectification ratio, and reverse voltage breakdown of 1.8V, 10^6 (±2V), and 390V, respectively was obtained for a 300μm diameter device at room temperature. Our results show remarkable stability of the I-V characteristics for 140-hour operation for devices operated continuously at 600°C and repeated thermal cycling (30 times) between 25-550°C in flowing Ar.

CONTROL ID: 3985717

CURRENT SYMPOSIUM: S4 - Complex Oxide Thin Films and Heterostructures: From Synthesis to Strain/Interface-engineered Emergent Properties

CURRENT SESSION: Strain, microstructures and functionality tuning in superlattices, vertical nanocomposites, and high entropy complex oxides

TITLE: ZnO-Au Hybrid Metamaterials Thin Films with Tunable Optical Properties

PRESENTATION TYPE: Contributed (Oral)

PRESENTER (FIRST NAME ONLY): Nirali

PRESENTER (LAST NAME ONLY): Bhatt

PRESENTER (INSTITUTION ONLY): Purdue University

PRESENTER (EMAIL ONLY): bhatt8@purdue.edu

PRESENTER (COUNTRY ONLY): United States

DATE/TIME SUBMITTED: September 25, 2023, 01:48 PM

ABSTRACT BODY:

Abstract Body: ZnO-Au nanocomposite thin films have been previously reported as hybrid metamaterials with unique optical properties such as plasmonic resonance properties and hyperbolic behaviors. In this talk, a discussion on how Au strain compensation has been effectively tuned by target composition variation and thus resulted in microstructure and optical properties tuning will be conducted. Specifically, all the ZnO-Au nanocomposite thin films grown through the pulsed laser deposition (PLD) method show obvious vertically aligned nanocomposite (VAN) structure with the Au nanopillars embedded in the ZnO matrix. Moreover, the average diameter of Au nanopillars increases as Au concentration increases, which also leads to the red shifts in the surface plasmon resonance (SPR) wavelength and changes in the hyperbolic behaviors of the films. As a whole, this talk will discuss how strain-driven tuning of optical properties and microstructure resulted through a novel Au concentration variation approach which has not been previously attempted in the ZnO-Au thin film system.

CONTROL ID: 3985718
CURRENT SYMPOSIUM: S14 - Emerging Semiconductor Materials and Interfaces
CURRENT SESSION: WBG and UWBG materials - synthesis, characterization, theory
TITLE: Structural Properties and Surface Morphology of III-N Thin Films Prepared via Sputter Deposition
PRESENTATION TYPE: Contributed (Oral)
PRESENTER (FIRST NAME ONLY): Joshua
PRESENTER (LAST NAME ONLY): Nordlander
PRESENTER (INSTITUTION ONLY): Penn-State University
PRESENTER (EMAIL ONLY): jln96@psu.edu
PRESENTER (COUNTRY ONLY): United States
DATE/TIME SUBMITTED: September 26, 2023, 01:57 PM
ABSTRACT BODY:
Abstract Body: In this presentation, we demonstrate that reactive High-Power Impulse Magnetron Sputtering (HiPIMS) is an effective alternative for depositing high quality, epitaxial III-N thin films. In contrast to conventional direct current (DC) sputtering, HiPIMS provides the needed kinetic energy and ionization fraction to establish a sufficiently reactive environment to promote full nitridation. Specifically, the low duty cycle regime of HiPIMS provides access to kW/cm^2 peak power densities without target degradation and thus produces dramatically increased metal reactivity.

This unique capability set enables us to prepare high crystal quality epitaxial InN thin films with low out-of-plane mosaicity and electron mobilities in excess of 400 cm^2/Vs with a step-and-terrace microstructure when deposited on AlN-nucleated sapphire substrates. Equilibrium supersaturation models are useful for predicting the surface microstructure of these films by varying the V/III ratio or deposition temperature.

In addition, epitaxial GaN thin films can be deposited on sapphire or on native GaN substrates with smooth surface morphologies characterized by c/2 steps and terraces. The presentation will discuss the relationships between sputtering parameters and III-N thin film crystal quality, surface morphology, and growth rate. Preliminary transport property measurements will be discussed.

CONTROL ID: 3985721

CURRENT SYMPOSIUM: S12 - Magnetic and Superconducting Materials: From Basic Science to Applications

CURRENT SESSION: Superconducting and related magnetic materials

TITLE: Multiple Intercalation Stages and Universal Superconducting Tc Enhancement through Polar Organic Species Intercalation into van der Waals Materials

PRESENTATION TYPE: Invited (by Invitation Only)

PRESENTER (FIRST NAME ONLY): Bing

PRESENTER (LAST NAME ONLY): Lv

PRESENTER (INSTITUTION ONLY): University of Texas at Dallas

PRESENTER (EMAIL ONLY): blv@utdallas.edu

PRESENTER (COUNTRY ONLY): United States

DATE/TIME SUBMITTED: September 25, 2023, 01:11 PM

ABSTRACT BODY:

Abstract Body: Chemical intercalation through inorganic ions such as alkali, alkaline-earth, and transition metals into van der Waals interlayer space, have been proven to be an effective method to tune structural, electronic, optical, and magnetic properties of many layered materials. Intriguingly, there are less theoretical and experimental studies on the intercalation effects of the "charge neutral" organic molecules into these materials. herein, we are using 1T-$SnSe_2$ as a model system to explore such effects, where we clearly demonstrated a universal superconducting Tc and superconducting volume fraction enhancement with polar organic species co-intercalation in comparison with solely inorganic Li ion intercalation. In addition, we also observe multiple intercalation stages with an intermediate stage clearly showing CDW-like transport anomalies. We have further extended this method to many other polar organic molecules and other van der Waals including 2D magnets and topological insulators. These results will be presented, and their implications will be discussed.

CONTROL ID: 3985752
CURRENT SYMPOSIUM: S2 - Advanced Microelectronics
CURRENT SESSION: Memristive switching and performance
TITLE: Volatile and Non-Volatile Resistive Switching in VO_2/La0.7Sr0.3MnO3/Al_2O_3 (0001) for Neuromorphic Computing
PRESENTATION TYPE: Contributed (Oral)
PRESENTER (FIRST NAME ONLY): Sundar
PRESENTER (LAST NAME ONLY): Kunwar
PRESENTER (INSTITUTION ONLY): Los Alamos National Lab
PRESENTER (EMAIL ONLY): sundar@lanl.gov
PRESENTER (COUNTRY ONLY): United States
DATE/TIME SUBMITTED: September 25, 2023, 01:43 PM
ABSTRACT BODY:
Abstract Body: Coexistence of non-volatile and volatile resistive switching (RS) modes in a single memristive device provides the flexibility to control the device for artificial neurons and synapses, enabling the creation of more energy and area efficient artificial neural networks for neuromorphic computing. In this talk, I will discuss our recent efforts on developing such devices that works as a volatile (neurons) and non-volatile (synapse) memory based on the control of biasing voltage. For this, we developed a VO_2/La0.7Sr0.3MnO3 (LSMO) heterostructure by pulsed laser deposition (PLD) and exploited its RS properties. Specifically, the non-volatile RS is found to be driven by the movement of oxygen vacancies at the VO_2/LSMO interface and requires high biasing voltage whereas the volatile RS is due to the MIT phase transition of VO_2 by Joule heating at low biasing voltage. Both RS modes are found to be stable, repeatable and can be reconfigured depending upon the magnitude of electric field and thus holds great promise for the application in memristive neural networks and neuromorphic computing.

CONTROL ID: 3985772
CURRENT SYMPOSIUM: S12 - Magnetic and Superconducting Materials: From Basic Science to Applications
CURRENT SESSION: Other magnetic materials
TITLE: Observation of mode-selective spin-phonon coupling in 2D antiferromagnets
PRESENTATION TYPE: Invited (by Invitation Only)
PRESENTER (FIRST NAME ONLY): Rahul
PRESENTER (LAST NAME ONLY): Rao
PRESENTER (INSTITUTION ONLY): Air Force Research Lab
PRESENTER (EMAIL ONLY): rahul.rao@wpafb.af.mil
PRESENTER (COUNTRY ONLY): United States
DATE/TIME SUBMITTED: September 25, 2023, 01:59 PM
ABSTRACT BODY:
Abstract Body: Two-dimensional magnetic materials offer the opportunity to study and manipulate emergent collective excitations. Among these, spin-phonon coupling is one of the most important interactions correlating charge, spin and lattice vibrations. Understanding and controlling this coupling is important for spintronics applications, control of magnons and phonon by THz radiation, and for strain-driven magnetoelastic applications. We report the observation of mode-selective spin-phonon coupling in several magnetic 2D metal thiophosphates ($NiPS_3$, $FePS_3$, $CoPS_3$ and $MnPS_3$) through multi-excitation and temperature-dependent resonant Raman scattering measurements. The phonon mode, which is a Raman-active out-of-plane vibrational mode (~ 250 cm^{-1} or 7.5 THz), exhibits an asymmetric Fano lineshape with its asymmetry proportional to the spin-phonon coupling. Ou measurements reveal the coupling to be the highest in $NiPS_3$, followed by $FePS_3$ and $CoPS_3$, and least in $MnPS_3$. The origin of the spin-phonon coupling is related to the unique structures of the 2D metal thiophosphates. Finally, the spin-phonon coupling is also observed in exfoliated materials, with a reduction by ~50% in the thinnest flakes compared to the bulk crystals.

CONTROL ID: 3985780
CURRENT SYMPOSIUM: S10 - Fundamental mechanisms in materials for next generation computing devices
CURRENT SESSION: Advanced characterization techniques
TITLE: Deterministic activation of resistive switching and light emission from TaOx with nanoscale resolution
PRESENTATION TYPE: Contributed (Oral)
PRESENTER (FIRST NAME ONLY): Olha
PRESENTER (LAST NAME ONLY): Popova
PRESENTER (INSTITUTION ONLY): Oak Ridge National Lab
PRESENTER (EMAIL ONLY): popovao@ornl.gov
PRESENTER (COUNTRY ONLY): United States
DATE/TIME SUBMITTED: September 25, 2023, 03:48 PM
ABSTRACT BODY:
Abstract Body: We have advanced both the measurement and the conditioning of the resistive switching phenomena utilizing amorphous thin TaOx films as a model system. We introduced 2 microscopy techniques –sMIM and CL- they can directly probe the electronic structure and conductance of amorphous binary oxide materials down with nanoscale resolution. Therefore, more robust connection to theoretical methods established, that can guide a design approach material optimization. We utilized ion-beam irradiation as a path to activate conductivity of materials without the need for energy-intensive electroforming. We demonstrate: (1) direct imaging of electronic structure and trap centers in TaOx; (2) electronic, dielectric and conductive characterization of TaOx after He-ion irradiation. Although ions modify the nature of bond centers at the lowest doses, films exhibit remarkable stability toward large displacive dose and are driven to metallic state at the limit of structural decomposition; (3) unexpected enhancement of luminescence yield with increasing damage of the film. Our techniques are directly compatible with device architectures, potentially allowing for a high-degree of on-demand tuning of neuromorphic circuitry and materials.

CONTROL ID: 3085785

CURRENT SYMPOSIUM: S4 - Complex Oxide Thin Films and Heterostructures: From Synthesis to Strain/Interface-engineered Emergent Properties

CURRENT SESSION: Phenomena arising from strain couplings and interface couplings including quantum phases and topological states and their control via external stimuli

TITLE: Hall anomaly driven by magnetic exchange interactions

PRESENTATION TYPE: Contributed (Oral)

PRESENTER (FIRST NAME ONLY): Pinku

PRESENTER (LAST NAME ONLY): Roy

PRESENTER (INSTITUTION ONLY): Los Alamos National Lab

PRESENTER (EMAIL ONLY): pinkur@lanl.gov

PRESENTER (COUNTRY ONLY): United States

DATE/TIME SUBMITTED: September 25, 2023, 02:17 PM

ABSTRACT BODY:

Abstract Body: Magnetic exchange interaction and broken inversion symmetry at the interface of two magnetic materials can give rise to a variety of magnetic behaviors. The study of magnetic properties in a heterostructure of two different magnetic materials reveals that exchange interaction modifies the magnetic behavior and helps in establishing a magnetic phase with two different anisotropies. The experimental Hall resistivity in such a modified magnetic phase exhibits temperature-dependent asymmetric hump and dip features. Our investigation reveals that the entanglement of two different Hall effects is a possible reason for such an anomaly.

CONTROL ID: 3985789
CURRENT SYMPOSIUM: S1: Characterization of Structure-Property Relationships in Functional Ceramics
CURRENT SESSION: Advances in scattering, imaging, and analytical techniques
TITLE: Local disorder in compositionally complex triple conducting oxides with neutron total scattering
PRESENTATION TYPE: Invited (by Invitation Only)
PRESENTER (FIRST NAME ONLY): Bernadette
PRESENTER (LAST NAME ONLY): Cladek
PRESENTER (INSTITUTION ONLY): University of Tennessee
PRESENTER (EMAIL ONLY): bcladek@vols.utk.edu
PRESENTER (COUNTRY ONLY): United States
DATE/TIME SUBMITTED: October 09, 2023, 12:32 PM
ABSTRACT BODY:
Abstract Body: Triple conducting oxides (TCOs), which simultaneously transport protons, oxygen ions, and electronic charge carriers, can enhance the performance and functionality of electrochemical energy conversion technologies. $BaCo_xFe_{0.8-x}Zr_{0.1}Y_{0.1}O_{3-d}$ (BCFZY, 0≤x≤0.7) is a compositionally complex rare-earth and transition metal co-doped alkaline earth TCO. Fundamental knowledge of the interplay of dopants on local structure, defect concentration, magnetism, and transport properties in BCFZY is limited. Here we explore the long-range and local structure of TCOs with varyingCo/Fe content using neutron diffraction and pair distribution function (PDF) studies performed from 10-500 K on the NOMAD and POWGEN beamlines at the Spallation Neutron Source, Oak Ridge National Laboratory. All compositions exhibit a high degree of local atomic distortion related to complex oxidation states. Quantitative analysis of diffraction and PDF data requires combined nuclear and magnetic Rietveld refinements, small-box, and large-box PDF modeling due to the complex composition, disorder, and magnetic structure of BCFZYs. Understanding the influence of cation composition on the local structure, carrier stability, and concentration of protonic defects will enable rational design and optimization of TCOs, tuning compositions for maximized performance.

CONTROL ID: 3985796

CURRENT SYMPOSIUM: S2 - Advanced Microelectronics

CURRENT SESSION: Dielectrics and ferroelectrics for device applications

TITLE: Anisotropic properties of epitaxial ferroelectric lead-free $0.5[Ba(Ti_{0.8}Zr_{0.2})O_3]$-$0.5(Ba_{0.7}Ca_{0.3})TiO_3$ films

PRESENTATION TYPE: Contributed (Oral)

PRESENTER (FIRST NAME ONLY): Nicholas

PRESENTER (LAST NAME ONLY): Cucciniello

PRESENTER (INSTITUTION ONLY): SUNY University at Buffalo

PRESENTER (EMAIL ONLY): ngcuccin@buffalo.edu

PRESENTER (COUNTRY ONLY): United States

DATE/TIME SUBMITTED: September 25, 2023, 02:34 PM

ABSTRACT BODY:

Abstract Body: As the energy demand is expected to double over the next 30 years, there has been a major initiative towards advancing the technology of both energy harvesting and storage for renewable energy. In this work, we explore a subset class of dielectrics, for energy storage, since ferroelectrics offer a unique combination of characteristics needed for energy storage devices. We investigate ferroelectric lead-free $0.5[Ba(Ti_{0.8}Zr_{0.2})O_3]$-$0.5(Ba_{0.7}Ca_{0.3})TiO_3$ epitaxial thin films with different crystallographic orientations grown by pulsed laser deposition. We focus our attention on the influence of the crystallographic orientation on the microstructure, ferroelectric, and dielectric properties. Our results indicate an enhancement of the polarization and strong anisotropy in the dielectric response for the (001)-oriented film. The enhanced ferroelectric, energy storage, and dielectric behavior of the (001)-oriented film is explained by the coexistence of orthorhombic-tetragonal phase, where the disordered local structure is in its free energy minimum.

CONTROL ID: 3985823

CURRENT SYMPOSIUM: S4 - Complex Oxide Thin Films and Heterostructures: From Synthesis to Strain/Interface-engineered Emergent Properties

CURRENT SESSION: Strain- and interface-controlled device performance in energy harvesting, memories, sensors, etc.

TITLE: Impact of Duty Cycle on the Endurance of Hafnium Zirconium Oxide Thin Film Capacitors

PRESENTATION TYPE: Contributed (Oral)

PRESENTER (FIRST NAME ONLY): Megan

PRESENTER (LAST NAME ONLY): Lenox

PRESENTER (INSTITUTION ONLY): University of Virginia

PRESENTER (EMAIL ONLY): wuf6nc@virginia.edu

PRESENTER (COUNTRY ONLY): United States

DATE/TIME SUBMITTED: September 25, 2023, 03:16 PM

ABSTRACT BODY:

Abstract Body: The effect of electric field cycling duty cycle on hafnium zirconium oxide (HZO) thin film capacitor endurance has been investigated. Capacitors were prepared comprising 10 nm of HZO deposited using plasma-enhanced atomic layer deposition with tungsten top and bottom electrodes. Grazing-incidence X-ray diffraction showed that the ferroelectric orthorhombic $Pca2_1$ HZO phase was present. 2 MV/cm field 10 kHz frequency bipolar pulse waves with 4%, 10%, 20%, 40%, and 100% effective duty cycles were used to cycle devices from a pristine unswitched state through wake-up and into fatigue. Polarization hysteresis, switching current, and PUND measurements were taken after each cycling interval. Results showed that increasing the duty cycle decreased capacitor fatigue lifetime, with a 4% duty cycle allowing a device to have a lifespan of over 10^8 cycles while a 100% duty cycles results in only 10^5 cycles before significant electrical leakage and failure. Increasing duty cycles results in increased wake up behavior at the expense of greater leakage currents, likely resulting from oxygen vacancy migration to interfaces. It is hypothesized that oxygen vacancy migration altered the energy barrier, promoting conduction and premature dielectric breakdown, with efforts to identify the degradation mechanism discussed.

CONTROL ID: 3985826

CURRENT SYMPOSIUM: S4 - Complex Oxide Thin Films and Heterostructures: From Synthesis to Strain/Interface-engineered Emergent Properties

CURRENT SESSION: Phenomena arising from strain couplings and interface couplings including quantum phases and topological states and their control via external stimuli

TITLE: Robust vortex state in a two-dimensional superconducting oxide heterostructure

PRESENTATION TYPE: Contributed (Oral)

PRESENTER (FIRST NAME ONLY): Ludi

PRESENTER (LAST NAME ONLY): Miao

PRESENTER (INSTITUTION ONLY): New Mexico State University

PRESENTER (EMAIL ONLY): lmiao@nmsu.edu

PRESENTER (COUNTRY ONLY): United States

DATE/TIME SUBMITTED: September 25, 2023, 05:18 PM

ABSTRACT BODY:

Abstract Body: Among unconventional superconductors, Sr_2RuO_4 (SRO) has attracted tremendous attention due to the potential application in quantum computing. A lot of mysteries, including order parameters are still not resolved. One of the reasons is that SRO is notoriously sensitive to even the smallest amount of impurities and dislations. For example, even a slight surface reconstruction of SRO is suffience to suppress the superconducting state, imposing great challenges to surface sensitive studies such as STM and ARPES to reveal its order parameter. For the same reason, a two-dimensional (2D) state in SRO is not accessible so far. In this work, we grow Ba:SRO/SRO/Ba:SRO heterostructures and study their magnetotransport at low temperatures. We accessed a 2D superconducting state in these heterostrucures. The minimal Ba doping almost does not affect crystal lattice or electronic structure of SRO above T_C, but locally breaks Coope pairs below T_C, leaving behind a thin layer of superfluid in the middle layer of the heterostructure. As a result, we have observed a resistive regime in the temperature-field phase diagram, due to the vortex liquid state, as a signature of the 2D superconductivity. In contrast to conventional 2D superconductors, the vortex liquid state in this case is very robust against magnetic field. We attribute such a robustness to the very large vortex sizes due to the unconventionality of SRO.

CONTROL ID: 3985828

CURRENT SYMPOSIUM: S3 - Advanced Electronic Materials: Processing and Structures and Properties and Applications

CURRENT SESSION: Advanced electronic materials, including ferroelectric, piezoelectric, dielectric, electrostrictive, electrocaloric and pyroelectric materials

TITLE: Coercive field and leakage current trends in Sc and B co-substituted AlN thin films

PRESENTATION TYPE: Contributed (Oral)

PRESENTER (FIRST NAME ONLY): John

PRESENTER (LAST NAME ONLY): Hayden

PRESENTER (INSTITUTION ONLY): Pennsylvania State University

PRESENTER (EMAIL ONLY): jvh5691@psu.edu

PRESENTER (COUNTRY ONLY): United States

DATE/TIME SUBMITTED: October 09, 2023, 07:48 PM

ABSTRACT BODY:

Abstract Body: The ferroelectric wurtzite nitrides AlScN and AlBN are of great research interest for their unique combination of large remanent polarizations, excellent retention, and back-end-of-the-line processing compatibility. The primary challenge for these materials lies in reducing film thicknesses to access relevant switching voltages while maintaining high leakage resistance. While AlScN films are attractive for the ability to tune coercive field by Sc composition modulation, AlBN films provide the advantage of lower leakage currents at high field, illuminating a potentially rich property space for the quaternary mixture. This work investigates the properties of scandium and boron co-substituted aluminum nitride thin films ($Al_{1-x-y}B_xSc_yN$) through first principles calculations and experiment. AlBScN thin films are deposited by magnetron co-sputtering using Al, BN, and Sc targets in a mixed argon and nitrogen atmosphere. Sputtered AlBScN films show reduced coercive field values compared to their AlBN counterparts, with increasing B concentrations causing a linear increase in coercive field values up to 7% BN. Current-voltage measurements show that addition of B to AlScN films reduces leakage currents while benefiting from a lower coercive field compared to AlBN. Additionally, fatigue and breakdown behavior are studied as a function of composition.

CONTROL ID: 3985839
CURRENT SYMPOSIUM: S9 - Machine Learning and Automated Synthesis/Characterization for Novel Materials
CURRENT SESSION: Computer-assisted experiment automation
TITLE: Automated Microscopy for Domain Wall Physics
PRESENTATION TYPE: Invited (by Invitation Only)
PRESENTER (FIRST NAME ONLY): Rama
PRESENTER (LAST NAME ONLY): Vasudevan
PRESENTER (INSTITUTION ONLY): Oak Ridge National Lab
PRESENTER (EMAIL ONLY): vasudevanrk@ornl.gov
PRESENTER (COUNTRY ONLY): United States
DATE/TIME SUBMITTED: September 25, 2023, 04:06 PM
ABSTRACT BODY:
Abstract Body: Ferroelectric domain walls can possess different functionality to the domains they separate and contribute significantly to static and dynamic material properties. At the same time, most studies of ferroelectric domains in ceramics have revolved around collections of domain walls, or isolated single domain walls studied at atomic resolution via electron microscopy. Although scanning probe microscopy (SPM) measurements exist at the mesoscale, these studies are often difficult to interpret, given the low numbers of statistics that is typically obtained.

In this work, we utilize automated experiments to collect a large number of images of ferroelectric domain walls in PbTiO3. First, we explored the relationship between the probing AC voltage applied to the SPM tip, and the observed width of the ferroelectric domain walls. We then wrote circular domains and acquired images at different excitation voltages to quantitatively determine the impact of the wall angle on the observed wall width.

Next, we observed the impact of surrounding domain wall structure on the depinning threshold via another automated experiment and observed a 'snapback' effect of the wall after bias pulsing. Physics-based autoencoders are the used to decipher the links between local domain wall structure and the domain wall properties. Implications for domain wall control via SPM and more advanced control schemes are discussed.

CONTROL ID: 3985854
CURRENT SYMPOSIUM: S14 - Emerging Semiconductor Materials and Interfaces
CURRENT SESSION: WBG and UWBG materials - synthesis, characterization, theory
TITLE: Epitaxial growth and properties of novel UWBG $MgGe_2O_4$ and $In_2Ge_2O_7$ compounds
PRESENTATION TYPE: Poster
PRESENTER (FIRST NAME ONLY): Kingsley
PRESENTER (LAST NAME ONLY): Egbo
PRESENTER (INSTITUTION ONLY): National Renewable Energy Laboratory
PRESENTER (EMAIL ONLY): kingsley.egbo@nrel.gov
PRESENTER (COUNTRY ONLY): United States
DATE/TIME SUBMITTED: September 25, 2023, 05:24 PM
ABSTRACT BODY:
Abstract Body: Recent high throughput computational search predicted Mg_2GeO_4 and $In_2Ge_2O_7$ as potential ultrawide bandgap oxide candidates for high voltage device applications. Here, we explore the thin film growth and properties of these two novel ternary oxides by pulsed laser deposition at an average pulse repetition rate of 8Hz and laser energy of 300mJ on different single crystalline substrates. Strong O_2 partial pressure dependence on composition is observed for $In_2Ge_2O_7$ growth due to the ease of desorption of GeO suboxide at low O_2 and high substrate temperature regimes. Results show that epitaxial spinel-type Mg_2GeO_4(100) is obtained on ~1.8% lattice mismatch $MgAl_2O_4$(100) and MgO (100) substrates, and epitaxial $In_2Ge_2O_7$(220) is stabilized on Al_2O_3(00.6). Bandgap values of 4.0 eV and 5.1 eV for $In_2Ge_2O_7$ and Mg_2GeO_4 respectively are obtained from spectroscopic ellipsometry. Preliminary transport measurement for undoped thin films shows that Mg_2GeO_4 is semi-insulating while $In_2Ge_2O_7$ shows sheet resistance >1MΩ. The sheet resistance decreased remarkably for Zr-doped $In_2Ge_2O_7$ confirming the predicted defect chemistry results suggesting that Zr can act as a donor in $In_2Ge_2O_7$. These results provide the basis for further material exploration and characterization of this new set of ultra-wide bandgap ternary oxide materials and will pave the way for their potential integration in high-power devices.

CONTROL ID: 3985879
CURRENT SYMPOSIUM: S15 - Electronics Materials for 5G/6G Applications
CURRENT SESSION: 5G/6G Devices
TITLE: Electro-optic sampling of on-chip integrated photodiodes for THz signal generation
PRESENTATION TYPE: Contributed (Oral)
PRESENTER (FIRST NAME ONLY): Anna
PRESENTER (LAST NAME ONLY): Osella
PRESENTER (INSTITUTION ONLY): Colorado School of Mines
PRESENTER (EMAIL ONLY): osella@mines.edu
PRESENTER (COUNTRY ONLY): United States
DATE/TIME SUBMITTED: September 25, 2023, 04:13 PM
ABSTRACT BODY:
Abstract Body: The generation of high fidelity mmWave and THz signals is imperative for the development of 6G technologies, however creating these signals with low noise has proven difficult with microwave up-conversion methods. We present electro-optic sampling of integrated on-chip photodiode devices for low-noise, high power sinusoidal wave generation in the THz range. Tight integration of optoelectronics to the device of interest allows for localized generation of signals enabling low loss signal transfer, whereas conventional optoelectronic integration relies on lossy interconnects. Furthermore, optoelectronic integration enables more than just sinusoidal signals, but optically modulated arbitrary waveforms.

CONTROL ID: 3985906
CURRENT SYMPOSIUM: S12 - Magnetic and Superconducting Materials: From Basic Science to Applications
CURRENT SESSION: Other magnetic materials
TITLE: Magnon engineering in YIG-based heterostructures - from strong coupling to short wavelength modes
PRESENTATION TYPE: Invited (by Invitation Only)
PRESENTER (FIRST NAME ONLY): Wei
PRESENTER (LAST NAME ONLY): Zhang
PRESENTER (INSTITUTION ONLY): UNC Chapel Hill
PRESENTER (EMAIL ONLY): zhwei@unc.edu
PRESENTER (COUNTRY ONLY): United States
DATE/TIME SUBMITTED: September 25, 2023, 04:42 PM
ABSTRACT BODY:
Abstract Body: Magnetic insulators such as yttrium iron garnets (YIGs) are of paramount importance for spin-wave or magnonic devices due to their ultralow power dissipation, exotic magnon state, and coherent coupling to other wave excitations. Magnetic insulator heterostructures bestow superior structural and magnetic properties and house many novel properties due to their strong and engineerable exchange interactions between individual layers. In this talk, I will discuss several YIG-based heterostructures in the context of strong magnon-magnon coupling, which enable a variety of coherent coupling phenomena, such as the magnonic-induced transparency, short wavelength excitation and selectivity, and nonlinear magnonic couplings. Finally, I will move from the conventional metal-insulator systems to a recently reported, all-insulating system exhibiting strong magnon-magnon hybridization. The versatility of magnon-magnon coupling in such heterostructures can prepare tailored magnon modes for their further hybridization with other solid-state excitations, including but not limited to optical photon, phonon, and qubits. The work is supported by NSF-ECCS-2246254.

References:
1. D. D. Awschalom, et al., IEEE Trans. Quantum Engineering 2, 5500836 (2021).
2. Y. Xiong, et al., Phys. Rev. Applied 17, 044010 (2022).
3. J. Inman, et al., Phys. Rev. Applied 17, 044034 (2022).
4. J. Liu, et al., arXiv:2309.03116.

CONTROL ID: 3985927
CURRENT SYMPOSIUM: S13 - Frontiers in Ferroic oxides: Synthesis, Properties, and Applications
CURRENT SESSION: Multiscale theory and modeling of ferroicity
TITLE: Phase-field simulation of bulk photovoltaic effect and associated domain manipulation in ferroelectrics
PRESENTATION TYPE: Contributed (Oral)
PRESENTER (FIRST NAME ONLY): Yi-De
PRESENTER (LAST NAME ONLY): Liou
PRESENTER (INSTITUTION ONLY): The University of Texas at Arlington
PRESENTER (EMAIL ONLY): yide.liou@uta.edu
PRESENTER (COUNTRY ONLY): United States
DATE/TIME SUBMITTED: September 25, 2023, 07:12 PM
ABSTRACT BODY:
Abstract Body: Optical control of ferroelectricity remains a continuous objective in the pursuit of advanced optoferroic devices. Recent research has unveiled the optical approach can indirectly couple with ferroelectric polarization through thermal depolarization, photostriction, and magnetoelectric effect. However, reports on how the light-induced space charges affect the domain reconfiguration remain scarce. In this work, we study the bulk photovoltaic (BPV) effect in ferroelectric oxides, which causes photoexcited carrier separation under above-bandgap illumination due to their non-centrosymmetric nature. A phase-field model incorporating BPV effect has been developed, enabling a microscopic understanding of the interplay between domain and BPV currents at ultrafast timescales. Our results indicate the variation in BPV currents between adjacent domains leads to the generation of opposite charges at domain walls and enhancing the electric field up to 10^6 V/cm within each domain. This electric field can rotate the polarization and aid in the movement of domain walls at sub-picosecond scale, which can realize non-volatile domain evolutions assisted by an anisotropic biaxial strain. Our study offers a fundamental understanding of BPV effects in ferroelectrics and associated ultrafast domain manipulation, with potential applications in the development of opto-ferroelectric memory devices.

CONTROL ID: 3985942
CURRENT SYMPOSIUM: S7 - Defects and Transport in Ceramics
CURRENT SESSION: Impact of dislocations and grain boundaries on ceramic functional properties
TITLE: Impedance Spectroscopy of Li_2OHCl Antiperovskite Solid Electrolytes: Grains vs. Grain Boundaries and In-Situ Synthesis Monitoring
PRESENTATION TYPE: Contributed (Oral)
PRESENTER (FIRST NAME ONLY): Kai-Wei
PRESENTER (LAST NAME ONLY): Lan
PRESENTER (INSTITUTION ONLY): University of Illinois at Urbana-Champaign
PRESENTER (EMAIL ONLY): kwlan2@illinois.edu
PRESENTER (COUNTRY ONLY): United States
DATE/TIME SUBMITTED: September 25, 2023, 09:07 PM
ABSTRACT BODY:
Abstract Body: Solid-state batteries have been widely studied due to the need for better energy density and safety. The antiperovskite Li_2OHCl is a promising oxide-based SE that has desirable ductile properties similar to those of sulfide-based SEs and exhibits good electrochemical stability against Li metal anodes. However, the reported ionic conductivities vary widely, and owing to the difficulty of separating grain and grain boundary contributions, there is uncertainty regarding whether/when the total ionic conductivity of Li_2OHCl is limited by the grain or grain boundary resistance. In this work, we synthesized polycrystalline cubic and orthorhombic Li_2OHCl using solid-state processing. Cold-pressed and elevated-temperature-sintered samples were prepared and measured by AC impedance spectroscopy (AC-IS). A distribution of relaxation times (DRT) analysis and equivalent circuits were applied to deconvolve the electrical processes in the samples. The formation of cubic Li_2OHCl from its reagents during heating and annealing was also investigated through in-situ AC-IS, and the corresponding phase evolution during the reaction and sintering stages was identified by in-situ 2D XRD. These investigations have provided insights into the various contributions to the total conductivity of Li_2OHCl across different temperature regimes.

CONTROL ID: 3985965
CURRENT SYMPOSIUM: S3 - Advanced Electronic Materials: Processing and Structures and Properties and Applications
CURRENT SESSION: Materials design, new materials and structures, and their emerging applications
TITLE: Theory-Guided Discovery of Antiferroelectric Ceramics for Energy-Efficient Capacitors
PRESENTATION TYPE: Contributed (Oral)
PRESENTER (FIRST NAME ONLY): Xiaoli
PRESENTER (LAST NAME ONLY): Tan
PRESENTER (INSTITUTION ONLY): Iowa State Univ
PRESENTER (EMAIL ONLY): xtan@iastate.edu
PRESENTER (COUNTRY ONLY): United States
DATE/TIME SUBMITTED: September 25, 2023, 05:24 PM
ABSTRACT BODY:
Abstract Body: The geometric nonlinear theory of martensitic phase transition has been successful in guiding the discovery of supercompatible shape memory alloys having ultralow thermal hysteresis and extended fatigue lifetime. Antiferroelectric ceramics, through the electric field-induced antiferroelectric (AFE) – ferroelectric (FE) phase transitions, have shown great promise for high-energy-density capacitors. However, only 70~80% of the stored energy is released during discharge (a charge-discharge energy efficiency of 70~80%). In the present work, we apply the geometric nonlinear theory to $PbZrO_3$-based oxides to predict the reversibility of the AFE – FE transition by assessing the AFE/FE interfacial lattice-mismatch strain. We observe a good correlation of interfacial mismatch strain with electric hysteresis, hence, the energy efficiency of AFE capacitors. Guided by theory, we conduct high-throughput materials search and discover AFE compositions with a near-perfect charge-discharge energy efficiency (98.2%). The fatigue life of the AFE capacitor reaches 80 million charge-discharge cycles, an 80 times enhancement over AFE ceramics with a large electric hysteresis.

CONTROL ID: 3985997
CURRENT SYMPOSIUM: S18 - Structure and function in high-entropy, entropy-stabilized, and related oxides
CURRENT SESSION: Characterization and quantification of structure and properties
TITLE: Designing frustration and metastable magnetism in single crystal compositionally complex oxides
PRESENTATION TYPE: Invited (by Invitation Only)
PRESENTER (FIRST NAME ONLY): Thomas
PRESENTER (LAST NAME ONLY): Ward
PRESENTER (INSTITUTION ONLY): Oak Ridge National Lab
PRESENTER (EMAIL ONLY): wardtz@ornl.gov
PRESENTER (COUNTRY ONLY): United States
DATE/TIME SUBMITTED: September 25, 2023, 05:58 PM
ABSTRACT BODY:

Abstract Body:

If we confine ourselves to binary and ternary materials, the limited number of spin active elements and the discrete nature of the spin (S) and magnetic exchange (J) parameters associated with these building blocks restrict the precision with which calculated states can be synthesized in the real world. High entropy and compositionally complex materials allow us to chaotically populate a crystal lattice with a mix of magnetically active elements that make it possible to access continuous tunability of magnetic parameter spaces. I will describe our recent work related to disorder effects across a range of strongly correlated magnetic systems. Emphasis will be given to observations that frustration can be driven by exchange disorder or variance in ionic radii to generate dynamic magnetic behaviors across a wide temperature range.

CONTROL ID: 3986006

CURRENT SYMPOSIUM: S11 - In situ/operando Characterization of Nanomaterials

CURRENT SESSION: Advances in developing in situ characterization techniques related to electron microscopy, X-ray, and neutron characterization techniques

TITLE: Operando dynamical studies of intercalated 2d materials via x-ray and electron scattering

PRESENTATION TYPE: Invited (by Invitation Only)

PRESENTER (FIRST NAME ONLY): Aaron

PRESENTER (LAST NAME ONLY): Lindenberg

PRESENTER (INSTITUTION ONLY): Stanford University

PRESENTER (EMAIL ONLY): aaronl@slac.stanford.edu

PRESENTER (COUNTRY ONLY): United States

DATE/TIME SUBMITTED: September 25, 2023, 06:05 PM

ABSTRACT BODY:

Abstract Body: I will describe two recent efforts focused on probing lithium-intercalation-driven metastable phases in the 2D transition metal dichalcogenide WTe_2 as measured by in-situ x-ray and electron scattering. In the first we find a new metastable phase featuring an anomalously large in-plane chemical expansion coefficient (expansion per unit lithium inserted). It is common for 2D materials to expand in the out-of-plane direction upon intercalation, but large tunable uniaxial in-plane expansion is much more rare. The unusual actuation of Li_xWTe_2 is linked to the formation of a metastable crystallographic phase, potentially enabling novel means for strain modulation at high frequencies [1]. In the second part of the talk I will describe new ways of probing the ultrafast dynamical response of this metastable phase, using a new setup which allows for pump-probe studies within an electrochemical cell. We find that Li-intercalation dramatically modulates electron-phonon coupling strengths in the material and provides new approaches for stabilizing novel phases of matter.

[1] Highly-efficient uniaxial in-plane stretching of a 2D material via ion insertion", P. Muscher, D. Rehn, A. Sood, K. Lim, D. Luo, X. Shen, M. Zajac, F. Lu, A. Mehta, Y. Li, X. Wang, E.J. Reed, W.C. Chueh, A.M. Lindenberg. Advanced Materials, 33, 2101875 (2021).

CONTROL ID: 3986017
CURRENT SYMPOSIUM: S13 - Frontiers in Ferroic oxides: Synthesis, Properties, and Applications
CURRENT SESSION: Advanced synthesis and characterization techniques for ferroic oxides
TITLE: Magnetic Scattering and Spectroscopy at High Pressures at APS and APS-U
PRESENTATION TYPE: Invited (by Invitation Only)
PRESENTER (FIRST NAME ONLY): Daniel
PRESENTER (LAST NAME ONLY): Haskel
PRESENTER (INSTITUTION ONLY): Argonne National Lab
PRESENTER (EMAIL ONLY): haskel@anl.gov
PRESENTER (COUNTRY ONLY): United States
DATE/TIME SUBMITTED: September 25, 2023, 06:21 PM
ABSTRACT BODY:
Abstract Body: Over the past two decades, developments at 3rd generation synchrotron light sources enabled x-ray studies of magnetism at high-pressure utilizing diamond anvil cells, under simultaneous application of high magnetic fields and low temperatures. Techniques such as x-ray resonant magnetic scattering and x-ray magnetic circular dichroism (XMCD) leveraged high flux density and energy/polarization tunability to provide unique insights into competing interactions in correlated electron systems. APS-U, a 4th generation synchrotron light source to come online in 2024, will provide increases of two orders of magnitude in coherence fraction and flux density of sub-micron x-ray beams. This allows extending high-pressure x-ray probes to the limit of static high-pressure generation, as well as enables in-situ imaging of electronic and phase inhomogeneity at near nanoscale dimensions using techniques such as resonant dichroic ptychography and real-space XMCD tomography. This talk provides an overview of previous developments at APS and upcoming opportunities at APS-U beamline POLAR.

CONTROL ID: 3986018

CURRENT SYMPOSIUM: S3 - Advanced Electronic Materials: Processing and Structures and Properties and Applications

CURRENT SESSION: Advanced electronic materials, including ferroelectric, piezoelectric, dielectric, electrostrictive, electrocaloric and pyroelectric materials

TITLE: Pressure and Electric Field Induced Domain Switching Mechanisms for Epitaxial & Polycrystalline Ferroelectric Thin films

PRESENTATION TYPE: Contributed (Oral)

PRESENTER (FIRST NAME ONLY): K M Abu Hurayra

PRESENTER (LAST NAME ONLY): Lizu

PRESENTER (INSTITUTION ONLY): University of Connecticut

PRESENTER (EMAIL ONLY): hurayralizu@uconn.edu

PRESENTER (COUNTRY ONLY): United States

DATE/TIME SUBMITTED: September 25, 2023, 09:05 PM

ABSTRACT BODY:

Abstract Body: Piezoresponse Force Microscopy (PFM) has emerged as a powerful tool with exceptional nanoscale resolution for the characterization of ferroelectric materials. In this study, we investigate different domain switching mechanisms in epitaxial La-doped BFO and polycrystalline PZT thin films in response to the application of force (stress) and bias, respectively. Employing simultaneous two-vector PFM enables continuous imaging of out-of-plane (normal) & in-plane (lateral) domain orientations for real-time mapping of domain nucleation and growth down to the nanoscale. Comparative analyses on nucleation sites, growth velocities of domain walls, and other aspects associated with switching phenomena elucidate fundamental differences in processing-structure-property relations for these samples. Domain switching and stability in ferroelectrics is of particular interest for filters, multi-layer chip capacitors, memory storage, and beyond CMOS technologies, with potential impacts on next-generation and quantum computing, IOT systems, and smart energy solutions.

CONTROL ID: 3986036
CURRENT SYMPOSIUM: S12 - Magnetic and Superconducting Materials: From Basic Science to Applications
CURRENT SESSION: Other magnetic materials
TITLE: Synthesis of new nitride material manganese cobalt nitride by RF co-sputtering
PRESENTATION TYPE: Contributed (Oral)
PRESENTER (FIRST NAME ONLY): Sita
PRESENTER (LAST NAME ONLY): Dugu
PRESENTER (INSTITUTION ONLY): National Renewable Energy Laboratory
PRESENTER (EMAIL ONLY): sita.dugu@gmail.com
PRESENTER (COUNTRY ONLY): United States
DATE/TIME SUBMITTED: September 25, 2023, 11:21 PM
ABSTRACT BODY:
Abstract Body: Transition metal(TM) nitrides have potential in industrial applications due to outstanding properties such as high hardness, thermal conductivity, and unique electrochemical properties. In 2019, Sun et al constructed a stability map of inorganic ternary metal nitrides that predicted hundreds of new ternary nitrides, out of which $MnCoN_2$ is the only TM_1–TM_2–N. Here series of Mn-Co-N films are synthesized and studied structural and magnetic properties.
Combinatorial MnCoN2 thin films are deposited using reactive cosputtering at temperatures range(25–450°C). Phases of the films is checked by XRD. Survey density function theory total energy calculations are performed in parallel on six prototype structures based on zincblende(zb) or rocksalt(rs) lattices. Comparing the calculated structures with experimental diffraction, films better match rs structures, also exhibit lower formation energy. However, only the primary diffraction peaks are seen, suggesting a large amount of cation antisite disorder. Mn and Co concentrations are measured by XRF and N concentration by RBS, that confirms 1:1:2 of Mn:Co:N with some reduced N due to O. Magnetic properties studied by SQUID demonstrates a weak moment of remanant magnetization 0.01 emu/gm and coercive field of 0.5 T and Tc ~10 K. To better understand the origin of the magnetic moment and its correlation with structure and stoichiometry, we perform XAS.

CONTROL ID: 3986041
CURRENT SYMPOSIUM: S8 - Data-driven and model-supported structure-property relationships in complex electroceramics
CURRENT SESSION: Local structure and properties of complex electroceramics
TITLE: Revisiting the Impact of Doping and Sample Thickness on the Electric-Field Induced Strain in Lead-Free $(K_{0.5}Na_{0.5})NbO_3$-based ceramics
PRESENTATION TYPE: Contributed (Oral)
PRESENTER (FIRST NAME ONLY): Muhammad
PRESENTER (LAST NAME ONLY): Wasim
PRESENTER (INSTITUTION ONLY): Sheffield Hallam University
PRESENTER (EMAIL ONLY): c2068528@hallam.shu.ac.uk
PRESENTER (COUNTRY ONLY): United Kingdom
DATE/TIME SUBMITTED: October 09, 2023, 03:08 PM
ABSTRACT BODY:
Abstract Body: $(K_{0.5}Na_{0.5})NbO_3$-based ceramics are among the most studied Pb-free piezoceramics, as potential candidates to replace piezoelectric actuators based on hazardous $Pb(Zr,Ti)O_3$ ceramics. For these applications, the magnitude and the temperature stability of the electric field induced strain and polarization are of utmost importance. The research efforts ranging from chemical doping of $(K_{0.5}Na_{0.5})NbO_3$ (KNN) ceramics to microstructural texturing have been explored over the last two decades as different approaches to enhance the piezoelectric performance of KNN ceramics. Recently, "giant" values of field induced strain exceeding 1%, have been reported for Sr-doped KNN ceramics. These anomalously large values were initially ascribed to the interaction between defect dipoles and domain switching. This prompted others to investigate a so-called thickness effect. Here, we are revisiting and establishing structure-property relationships and we also investigate samples of different thicknesses. In addition, we evaluate and compare the impact of incorporating other divalent species, including Ca and Ba, into the A-site of $(KNa)NbO_3$-based ceramics. Our investigation involves a comparative analysis of the electric field-induced polarization and strain characteristics of these ceramics, shedding light on effects of atomic radii variations, thicknesses and poling.

CONTROL ID: 3986056
CURRENT SYMPOSIUM: S7 - Defects and Transport in Ceramics
CURRENT SESSION: Structure and stability of defects and defect complexes via in-situ measurement
TITLE: Toward near-zero chemical strains in mixed- and proton-conducting perovskites by tailoring bond architecture
PRESENTATION TYPE: Contributed (Oral)
PRESENTER (FIRST NAME ONLY): Nicola
PRESENTER (LAST NAME ONLY): Perry
PRESENTER (INSTITUTION ONLY): University of Illinois Urbana-Champaign
PRESENTER (EMAIL ONLY): nhperry@illinois.edu
PRESENTER (COUNTRY ONLY): United States
DATE/TIME SUBMITTED: September 25, 2023, 08:07 PM
ABSTRACT BODY:
Abstract Body: Perovskite lattices expand when they lose oxygen or become hydrated, and this chemical strain can translate to performance degradation in constrained multilayer devices and composites. To prevent this failure route, we seek crystal chemical design rules for materials with near-zero coefficients of chemical expansion (CCEs). ABO_3 perovskites support a range of BO_6 octahedral tilt patterns, and the related B-O-B angles and charge distribution become bond architecture elements to tune in search of CCE control knobs. We evaluate isothermal CCEs by combining measurements of stoichiometry changes from thermogravimetry with strains from dilatometry in varying oxygen and steam partial pressures. Recently we have examined how different octahedral tilt patterns in proton-conducting $(Ba,Sr)(Ce,Zr,Y)O_{3-x}$ ceramics correlate to hydration CCEs, adding in-situ neutron diffraction for atomistic insight into hydrated vs. dry structures at elevated temperatures. We also examined the relationship between covalency and redox CCEs in $(La,Y,Ca)FeO_{3-x}$ compositions, adding X-ray absorption spectroscopy and synchrotron X-ray diffraction to understand charge distribution and bond angles. We find a minimum in CCE at relatively low tolerance factor (intermediate distortions), with some orthorhombic perovskites providing highly anisotropic lattice strains that result in low macroscopic strains.

CONTROL ID: 3986060

CURRENT SYMPOSIUM: S5 - Controlling Grain Boundary Structure, Chemistry, and their Network as a Function of Mtl Processing

CURRENT SESSION: Nanocrystalline ceramics

TITLE: Grain size dependence of mechanical properties of nanocrystalline magnesium aluminate $MgAl_2O_4$

PRESENTATION TYPE: Invited (by Invitation Only)

PRESENTER (FIRST NAME ONLY): Seok-Woo

PRESENTER (LAST NAME ONLY): Lee

PRESENTER (INSTITUTION ONLY): University of Connecticut

PRESENTER (EMAIL ONLY): seok-woo.lee@uconn.edu

PRESENTER (COUNTRY ONLY): United States

DATE/TIME SUBMITTED: September 25, 2023, 08:06 PM

ABSTRACT BODY:

Abstract Body: To develop transparent materials with superior mechanical properties, nanocrystalline magnesium aluminate ($MgAl_2O_4$) spinel with grain sizes ranging from 3.7 to 80nm has been synthesized by environmentally controlled pressure assisted sintering. In this study, we investigated the microstructure and grain size dependence of the mechanical properties of nanocrystalline $MgAl_2O_4$ by performing transmission electron microscopy, nanoindentation, uniaxial micropillar compression, and micro-cantilever bending. Electron microscopy confirmed that the environmentally controlled pressure assisted sintering technique produces a nearly fully dense grain structure with a porosity of less than 1% in larger grain-sized ceramics and pore-free grain structures in the smaller grain-sized ceramics. Mechanical characterization revealed that nanoindentation hardness, compressive fracture strength, and fracture toughness each exhibit distinct grain size dependence. Our results point to a change in dominant strain accommodating mechanisms from dislocation-based plasticity to grain boundary assisted plasticity as the grain size is reduced. Practical implications of the change in strain accommodation mechanisms manifest as the emergence of weak grain size dependence of hardness and strength, and a ~2-fold increase in apparent fracture toughness for the smaller grain-sized ceramics.

CONTROL ID: 3986083
CURRENT SYMPOSIUM: S12 - Magnetic and Superconducting Materials: From Basic Science to Applications
CURRENT SESSION: Superconducting and related magnetic materials
TITLE: Materials Research for Scaling Superconducting Quantum Electronics
PRESENTATION TYPE: Invited (by Invitation Only)
PRESENTER (FIRST NAME ONLY): Kasra
PRESENTER (LAST NAME ONLY): Sardashti
PRESENTER (INSTITUTION ONLY): Clemson University
PRESENTER (EMAIL ONLY): ksardas@g.clemson.edu
PRESENTER (COUNTRY ONLY): United States
DATE/TIME SUBMITTED: September 25, 2023, 08:33 PM
ABSTRACT BODY:
Abstract Body: Error-corrected quantum computing demands seamless integration of 10s of thousands of qubits. While superconducting qubits take the leading role in near-term quantum computing, their scaling is limited by the errors and losses in individual qubits as well as the size of the comprising microwave components, and the cryostat that houses them. This beckons extensive materials research to engineer platforms ideal for fault-tolerant and scalable quantum hardware. Here, I discuss hybrid superconductor-semiconductor (super-semi) materials systems as promising platforms for scalable quantum circuits by enabling dissipationless tuning of superconducting qubits and couplers. I will provide an overview of the advances and limitations of the state-of-the-art voltage-tunable superconducting quantum devices demonstrated on III-V heterostructures. I will then discuss a few strategies to break through the existing barriers for voltage-tunable devices by integrating the more robust group IV semiconductors into hybrid super-semi systems. Through this approach, I hope to emphasize the significant role materials physics plays in the development of emerging quantum hardware.

CONTROL ID: 3986084
CURRENT SYMPOSIUM: S7 - Defects and Transport in Ceramics
CURRENT SESSION: Structure and stability of defects and defect complexes via in-situ measurement
TITLE: Toward a Defect Chemical Model of $Y_2Ru_2O_7$ Pyrochlore, an OER Catalyst for Acidic Conditions
PRESENTATION TYPE: Poster
PRESENTER (FIRST NAME ONLY): En Ju
PRESENTER (LAST NAME ONLY): Cho
PRESENTER (INSTITUTION ONLY): University of Illinois at Urbana-Champaign
PRESENTER (EMAIL ONLY): enjucho2@illinois.edu
PRESENTER (COUNTRY ONLY): United States
DATE/TIME SUBMITTED: September 25, 2023, 09:18 PM
ABSTRACT BODY:
Abstract Body: Ru-containing pyrochlores ($A_2Ru_2O_{7-\delta}$) are known for good oxygen evolution reaction (OER) catalytic activity in acidic environments; however, the mechanism is not fully established. We seek to understand how the ionic and electronic defect chemistry and transport behavior may correlate to the electrochemical behavior, building on the possibility that lattice oxygen is involved in the mechanism and that Fermi level may be an important descriptor. To fill the mechanistic knowledge gap and establish design principles for high activity and stability, we perform the first in-depth defect chemical study of the rare acid-stable electrocatalysts. Custom-built thermogravimetric analysis with rapid gas exchange and ac impedance spectroscopy are applied to electrocatalyst nanoparticles as a function of oxygen partial pressure at various isotherms, and as a function of temperature at single atmospheres, to examine mass and conductivity changes. We first quantify room-temperature oxygen stoichiometry of the Y doped (0, 10%, 20%, 30%, 40%) $Y_2(Y_xRu_{2-x})O_{7-\delta}$ series to examine how the varied oxygen stoichiometry with different Y dopant content correlates with the catalytic performance. We then measure how equilibrium oxygen stoichiometry and conductivity of $Y_2Ru_2O_{7-\delta}$ changes with oxygen partial pressure and temperature, to build a defect chemical model. Implications for catalyst design will be addressed.

CONTROL ID: 3986088
CURRENT SYMPOSIUM: S18 - Structure and function in high-entropy, entropy-stabilized, and related oxides
CURRENT SESSION: Characterization and quantification of structure and properties
TITLE: Probing Local Structure Transitions in High Entropy Fluorite Oxides using XANES
PRESENTATION TYPE: Contributed (Oral)
PRESENTER (FIRST NAME ONLY): Gerald
PRESENTER (LAST NAME ONLY): Bejger
PRESENTER (INSTITUTION ONLY): Virginia Tech
PRESENTER (EMAIL ONLY): jerry.b43211@gmail.com
PRESENTER (COUNTRY ONLY): United States
DATE/TIME SUBMITTED: September 25, 2023, 09:07 PM
ABSTRACT BODY:
Abstract Body: Emerging complexities in rare-earth high entropy fluorite oxides are of current interest as they are promising for electro-mechanical applications and as ionic conductors. Ideal bixbyite is of the form A_2O_3 (Ia-3) where 3+ cations are 6 coordinated to oxygen, while fluorite (Fm-3m), of form AO2, maintains 4+ cations that are 8 coordinated to oxygen. Compositions $(LaCePrSmY)_{0.2}O_2$ and $Ce_x(LaPrSmY)_{1-x}O_2$ have been investigated to understand the transition from the bixbyite phase to that of pure fluorite from the local structure perspective of each cation using X-ray near edge fine structure (XANES). Through peak fitting of white line and pre-edge features of L3 and L1 absorption edges, respectively, estimations of the coordination and valence are discussed to help understand this transition from bixbyite to fluorite.

CONTROL ID: 3986103

CURRENT SYMPOSIUM: S7 - Defects and Transport in Ceramics

CURRENT SESSION: Defect mobility and transport behavior, including under operando or in extreme environments

TITLE: Low Thermal-Budget Processing of Thin-Film Li_2OHCl Antiperovskite Solid Electrolytes for All-Solid-State Batteries

PRESENTATION TYPE: Poster

PRESENTER (FIRST NAME ONLY): Kai-Wei

PRESENTER (LAST NAME ONLY): Lan

PRESENTER (INSTITUTION ONLY): University of Illinois at Urbana-Champaign

PRESENTER (EMAIL ONLY): kwlan2@illinois.edu

PRESENTER (COUNTRY ONLY): United States

DATE/TIME SUBMITTED: September 25, 2023, 09:41 PM

ABSTRACT BODY:

Abstract Body: Vapor-phase deposition may represent a suitable route for low-temperature electrolyte growth on thermally sensitive electrodes, enabling all-solid-state batteries with pristine and tailored interfaces. Lithium phosphorus oxygen nitrogen (LiPON) and lithium lanthanum zirconium oxide (LLZO) are extensively studied solid electrolyte films, where higher processing temperatures may promote structural rearrangement beneficial to conductivity but also detrimental to Li loss. In this work, to address the challenge of balancing lithium loss with sufficient thermal energy to crystallize, we investigated the thin-film growth of antiperovskite Li_2OHCl, with a much lower melting point (~300 C^{O}) than LLZO or LiPON. Owing to the paucity of reports on pulsed laser deposition of Li oxyhalides, we conducted a systematic study of the effect of deposition parameters on the structure and conductivity of Li_2OHCl films. Through X-ray diffraction, in-plane AC-impedance spectroscopy using microelectrodes, and nuclear reaction analysis, we related deposition temperature and laser repetition rate to the phase composition, crystalline quality, stoichiometry, and ultimately ionic conductivity. Our results contribute to understanding processing-structure-composition-property relationships of low-temperature-deposited, exquisitely air-sensitive Li_2OHCl films for all-solid-state batteries.

CONTROL ID: 3986105
CURRENT SYMPOSIUM: S13 - Frontiers in Ferroic oxides: Synthesis, Properties, and Applications
CURRENT SESSION: Ferroic materials for novel devices
TITLE: High-performance domain-engineered tunable dielectrics
PRESENTATION TYPE: Invited (by Invitation Only)
PRESENTER (FIRST NAME ONLY): Jonathan
PRESENTER (LAST NAME ONLY): Spanier
PRESENTER (INSTITUTION ONLY): Drexel University
PRESENTER (EMAIL ONLY): spanier@drexel.edu
PRESENTER (COUNTRY ONLY): United States
DATE/TIME SUBMITTED: September 25, 2023, 10:04 PM
ABSTRACT BODY:
Abstract Body: Realization of tunable materials that are multifunctional and maintain high performance in dynamically changing environments is a fundamental goal of materials science and engineering. Tunable dielectrics form the basis of a wide variety of communication and sensing devices and require breakthrough performance improvement to enable next-generation technologies. Engineering the ferroelectric polarization-energy landscape offers intriguing new opportunities for tailoring properties. I will present our recent collaborative work to design and model the domain structure-property relationships of film materials that exhibit exceptionally large tunable dielectric response over a wide temperature range and feature tunability that is itself controllable. These results suggest that domain engineering is a powerful approach for achieving unprecedented modulation of functional properties in ferroelectric films. Work supported by the US Army Research Office under grant no W911NF-21-1-0126 and the US Army Research Laboratory under W911NF-19-2-0119.

CONTROL ID: 3986109
CURRENT SYMPOSIUM: S10 - Fundamental mechanisms in materials for next generation computing devices
CURRENT SESSION: Advanced characterization techniques
TITLE: Data-driven dynamics of nanoscale hysteresis for systematic understanding of switching mechanisms in neuromorphic materials
PRESENTATION TYPE: Contributed (Oral)
PRESENTER (FIRST NAME ONLY): Petro
PRESENTER (LAST NAME ONLY): Maksymovych
PRESENTER (INSTITUTION ONLY): Oak Ridge National Lab
PRESENTER (EMAIL ONLY): 5nm@ornl.gov
PRESENTER (COUNTRY ONLY): United States
DATE/TIME SUBMITTED: September 25, 2023, 09:42 PM
ABSTRACT BODY:
Abstract Body: Unraveling the fundamental processes of device function is a long-standing challenge, which is exacerbated by the non-equilibrium nature of material dynamics relevant to computing. Scanning probe microscopy can bridge the nanoscale gap between idealized atomic models and the device scale. However, nanoscale measurements often suffers from lack of reliable quantification. Here I will present our recent work on robust inference of polarization dynamics in multiwell ferroelectrics, directly measuring the multiwell structure of the polarization potential in thiophosphate ferroelectrics [1-3]. I will discuss the topological data analysis of the multiwell signatures, utilzing dynamic embedding representations for effective comparison across measurements and reference to numerical models on dynamical grounds. I will also discuss the potential of multimodal microscopy to reveal still elusive intrinsic dynamics of electroresistive oxides. Research funded in part by the DOE Office of Science Research Program for Microelectronics Codesign through the Abisko Project and the Center for Nanophase Materials Sciences, Oak Ridge National Laboratory, which is a DOE sponsored User facility. 1. Neumayer/PM, Adv. Mat. 20 (2023) 2211194. 2. Neumayer/PM et al., ACS Appl. Mat. Int.14 (2022) 3018. 3. Neumayer/PM, Neuromorph. Comput. Eng. 3 (2023) 014005.

CONTROL ID: 3986133
CURRENT SYMPOSIUM: S12 - Magnetic and Superconducting Materials: From Basic Science to Applications
CURRENT SESSION: Superconducting and related magnetic materials
TITLE: Tunneling Andreev Reflection - new quantitative microscopy of superconductors with atomic resolution
PRESENTATION TYPE: Contributed (Oral)
PRESENTER (FIRST NAME ONLY): Petro
PRESENTER (LAST NAME ONLY): Maksymovych
PRESENTER (INSTITUTION ONLY): Oak Ridge National Lab
PRESENTER (EMAIL ONLY): 5nm@ornl.gov
PRESENTER (COUNTRY ONLY): United States
DATE/TIME SUBMITTED: September 25, 2023, 09:59 PM
ABSTRACT BODY:
Abstract Body: Renewed emphasis on unconventional superconductivity is driven by explosion of new materials and numerous proposals seeking exotic superconductivity for quantum information functions. In this talk, we will present our recent developments of tunneling Andreev reflection (TAR) - designed to detect and quantify Andreev reflection across atomic-scale tunnel junctions [1,2]. The most notable capability of TAR is distinguishing the number of Andreev reflections across the tunnel junction. In most cases TAR can detect at least two distinct Andreev measurements [1,2]. We find that contrast between specific reflections is particularly strong for unconventional superconductors, where suppression, enhancement and other spectral features emerge depending on the specific order parameter symmetry [2]. TAR therefore provides a unique approach to detect unconventional superconductivity and further explore the symmetry of the superconducting order parameter, with robust theory-experiment comparison. Research sponsored by Division of Materials Science and Engineering, BES, Office of Science, US DOE. SPM experiments carried out at the Center for Nanophase Materials Sciences, Oak Ridge National Laboratory, a US DOE Office of Science User Facility. 1. Nano Letters 23 (2023) 8310. 2. Nano Letters 22 (2022) 4042.

CONTROL ID: 3986140

CURRENT SYMPOSIUM: S4 - Complex Oxide Thin Films and Heterostructures: From Synthesis to Strain/Interface-engineered Emergent Properties

CURRENT SESSION: Strain, microstructures and functionality tuning in superlattices, vertical nanocomposites, and high entropy complex oxides

TITLE: Oxygen-vacancy-induced magnetic anisotropy in $LaCrO_3$/$LaMnO_3$ heterostructures

PRESENTATION TYPE: Contributed (Oral)

PRESENTER (FIRST NAME ONLY): Divine

PRESENTER (LAST NAME ONLY): Kumah

PRESENTER (INSTITUTION ONLY): Duke University

PRESENTER (EMAIL ONLY): dpkumah@ncsu.edu

PRESENTER (COUNTRY ONLY): United States

DATE/TIME SUBMITTED: September 25, 2023, 10:23 PM

ABSTRACT BODY:

Abstract Body: Structural, electronic, and spin interactions at the interfaces between atomically thin complex oxides provide a route to engineer the magnetic properties of a system with applications in spintronic devices. Here, we investigate the relationship between magnetic anisotropy, oxygen octahedral rotations, and the oxygen stoichiometry in $LaCrO_3$/$LaMnO_3$(LCO/LMO) superlattices grown by molecular beam epitaxy. High-resolution synchrotron X-ray diffraction-based imaging of the oxygen octahedral distortions reveals a change in the lattice symmetry on annealing in oxygen. The structural transition is correlated with in-plane magnetic anisotropy for as-grown oxygen-deficient films and isotropic magnetic moments along the principal axis for stoichiometric films. Magnetic ordering in this system arises due to an antiferromagnetic coupling between Mn and Cr across the LMO/LCO interface confirmed by X-ray magnetic circular dichroism measurements. Our finding leads to a better understanding of the microscopic origin of the magnetic anisotropy in this system and demonstrates a pathway to engineering the magnetic anisotropy with oxygen vacancies in transition metal oxide heterostructures.

This work is supported by the National Science Foundation under Grant No. NSF DMR-1751455

CONTROL ID: 3986143
CURRENT SYMPOSIUM: S2 - Advanced Microelectronics
CURRENT SESSION: Role of defects, interface and synthesis/fabrication process on device properties
TITLE: Ion Migration and Chemical Reactivity Studies in Electronic Materials by Means of Time-of-Flight Secondary Ion Mass Spectrometry
PRESENTATION TYPE: Invited (by Invitation Only)
PRESENTER (FIRST NAME ONLY): Anton
PRESENTER (LAST NAME ONLY): Ievlev
PRESENTER (INSTITUTION ONLY): Oak Ridge National Lab
PRESENTER (EMAIL ONLY): ievlevav@ornl.gov
PRESENTER (COUNTRY ONLY): United States
DATE/TIME SUBMITTED: September 25, 2023, 10:14 PM
ABSTRACT BODY:
Abstract Body: The performance of various electronic devices is defined by the delicate interplay of electrical response and charge carrier migration at the nanoscale. Although physical behavior and macroscopic functional response of these materials is well established, intrinsic chemical phenomena can dramatically alter their behavior and thus restrict area of utilization. Over the last decade, advancements in development of novel nanoscale characterization tools have revolutionized our understanding of material physical response; however, dynamic electrochemical behavior and ion migration remain poorly understood. Recently ToF-SIMS has proven to be effective tool for characterization of static chemical states in energy materials.

Here we introduce approach based on combined AFM/ToF-SIMS platform for correlated studies of the dynamic chemical phenomena on the nanoscale in operando conditions. Being used for characterization of the range of electronic materials, including ferroelectrics, photovoltaics and memristors it allowed direct observation of the ionic migration within the device under externally applied stimuli. Altogether, developed approaches enable direct characterization of interplay between chemical and functional response in variety of materials, which aids in the development and optimization of novel devices and applications.

CONTROL ID: 3986163
CURRENT SYMPOSIUM: S9 - Machine Learning and Automated Synthesis/Characterization for Novel Materials
CURRENT SESSION: Data-driven modeling and analysis
TITLE: Refining Solvation Free Energy Predictions in Vaspsol Through Optimization
PRESENTATION TYPE: Contributed (Oral)
PRESENTER (FIRST NAME ONLY): Sean
PRESENTER (LAST NAME ONLY): Florez
PRESENTER (INSTITUTION ONLY): University of Florida
PRESENTER (EMAIL ONLY): seanflorez@ufl.edu
PRESENTER (COUNTRY ONLY): United States
DATE/TIME SUBMITTED: October 04, 2023, 07:43 PM
ABSTRACT BODY:
Abstract Body: Density Functional Theory (DFT) is essential for predicting material properties and reaction barriers. However, computational costs increase when solvent molecules surround the system of interest. Continuum models like VASPsol approximate solvation effects using the Possion-Boltzmann model. This study improved VASPsol's accuracy by developing artificial neural networks to predict solvation energy error with the PBE functional. COSMO sigma profiles, computed with Gaussian09 and the cc-pvTZ basis setwere used as molecular descriptors. The predicted errors were input into a Nelder Mead optimization process to identify optimal parameters for minimizing VASPsol solvation energy errors. We evaluated the estimated optimized parameters using VASPsol and repeated this process to reduce errors. VASPsol's error was reduced from 1.17 kcal/mol to 1.05 kcal/mol for all 274 solutes solvated in water in the Truhlar set. Our work enhances the accuracy and reliability of VASPsol, contributing to the advancement of continuum models in DFT.

CONTROL ID: 3986167
CURRENT SYMPOSIUM: S12 - Magnetic and Superconducting Materials: From Basic Science to Applications
CURRENT SESSION: Other magnetic materials
TITLE: Magnetism of Rare Earth Nitride Perovskites
PRESENTATION TYPE: Contributed (Oral)
PRESENTER (FIRST NAME ONLY): Rebecca
PRESENTER (LAST NAME ONLY): Smaha
PRESENTER (INSTITUTION ONLY): National Renewable Energy Laboratory
PRESENTER (EMAIL ONLY): rebecca.smaha@nrel.gov
PRESENTER (COUNTRY ONLY): United States
DATE/TIME SUBMITTED: September 25, 2023, 11:38 PM
ABSTRACT BODY:
Abstract Body: Nitrides are a rich area of magnetic materials, but developments in nitrides have been stymied by the difficulty of synthesizing them, given the comparative stability of N_2. As such, the phase space of nitrides is vastly underexplored compared to those of oxides and halides. Recent advances in experimental methods for synthesizing nitride materials, particularly thin film methods employing radiofrequency co-sputtering to increase the effective synthesis temperature, have led to a rapid increase in the number of known ternary nitride phases. The greater covalency of nitrogen than oxygen and halogens makes nitrides attractive for semiconducting and other functional applications in renewable energy, and incorporating rare earth cations into novel ternary nitrides may enable novel functionality for magnetic, (opto)electronic, spintronic, and other applications. Rare earth nitride perovskites offer an intriguing path forward for multiferroics since there is evidence of piezoelectricity in $LaWN_3$ and low temperature magnetic order in $CeMoN_3$. I present an overview of the magnetic properties of the emerging family of rare earth nitride perovskites, which now number at least 5, and discuss the development of design rules for the stabilization and functionalization of new nitride perovskites.

CONTROL ID: 3986178
CURRENT SYMPOSIUM: S19 - Failure: The Greatest Teacher
CURRENT SESSION: None
TITLE: More inclusive Q&A will lead to better science
PRESENTATION TYPE: Poster
PRESENTER (FIRST NAME ONLY): Christopher
PRESENTER (LAST NAME ONLY): Rom
PRESENTER (INSTITUTION ONLY): National Renewable Energy Laboratory
PRESENTER (EMAIL ONLY): christopher.rom@nrel.gov
PRESENTER (COUNTRY ONLY): United States
DATE/TIME SUBMITTED: September 25, 2023, 11:45 PM
ABSTRACT BODY:
Abstract Body: Traditional question and answer (Q&A) sessions are a sub-optimal use of conference time because they do not engage the full audience. Since 2017, several studies in a range of academic disciplines have identified that women are under-represented among question-askers. Insufficiently inclusive Q&As leave thought-provoking questions un-asked, meaning our communities miss out on ideas worth hearing. Throughout my PhD and into my postdoc, I have collected data on question asking behavior at the conference talks I attended (ceramics and solid-state chemistry, broadly speaking), revealing that this disparity is present in our community too. Fortunately, prior research shows that merely discussing this trend can help balance Q&A sessions. In this poster, I will summarize existing research on gender equity in Q&A sessions, present my own data relevant to the EMA community, and describe preliminary results from possible alternative Q&A formats. This work will spark conversations around how to structure our scientific talks to better foster learning and collaboration, boosting our ability to conduct cutting edge science.

CONTROL ID: 3986193
CURRENT SYMPOSIUM: S2 - Advanced Microelectronics
CURRENT SESSION: Theory, modeling and first principles calculations of devices performance and properties
TITLE: New Polarization Switching Mechanisms in $Al_{1-x}M_xN$ (M = Sc, B, La, Gd) Alloys
PRESENTATION TYPE: Contributed (Oral)
PRESENTER (FIRST NAME ONLY): Cheng-Wei
PRESENTER (LAST NAME ONLY): Lee
PRESENTER (INSTITUTION ONLY): Colorado School of Mines
PRESENTER (EMAIL ONLY): clee2@mines.edu
PRESENTER (COUNTRY ONLY): United States
DATE/TIME SUBMITTED: September 25, 2023, 11:41 PM
ABSTRACT BODY:
Abstract Body: Wurtzite-type ferroelectrics have drawn increasing attention due to the promise of better performance and integration than traditional oxide ferroelectrics with semiconductors such as Si, SiC, and III-V compounds. However, wurtzite-type ferroelectrics generally require enormous electric fields, approaching breakdown, to reverse their polarization. The underlying switching mechanism(s) remain elusive, especially in AlN-based alloys. Here, we use first-principles computations to systematically understand the switching behavior of $Al_{1-x}M_xN$ alloys (M = Sc, B, La, Gd) as a function of the alloy composition. It is generally assumed that switching in these alloys proceed via the wurtzite-hexagonal-wurtzite pathway (collective mechanism). Interestingly, we find that the switching pathway changes from the commonly assumed collective mechanism to a lower-barrier mechanism enabled by inversion of individual tetrahedra (individual mechanism) with increased M composition. Therefore, the dominant switching mechanism depends on the alloy composition. We also find that larger structural distortion in AlN induced by alloying promotes individual switching mechanism and thus, lower switching barriers. Our findings provide insights for future engineering and realization of wurtzite-type ferroelectric materials and open a door to understanding domain motion.

CONTROL ID: 3986237

CURRENT SYMPOSIUM: S5 - Controlling Grain Boundary Structure, Chemistry, and their Network as a Function of Mtl Processing

CURRENT SESSION: Extreme temperatures (cold sintering, SPS, etc.)

TITLE: Grain Boundary Chemistry of Ceramics Fabricated by Selective Laser Sintering

PRESENTATION TYPE: Invited (by Invitation Only)

PRESENTER (FIRST NAME ONLY): Bai

PRESENTER (LAST NAME ONLY): Cui

PRESENTER (INSTITUTION ONLY): University of Nebraska–Lincoln

PRESENTER (EMAIL ONLY): bcui3@unl.edu

PRESENTER (COUNTRY ONLY): United States

DATE/TIME SUBMITTED: September 26, 2023, 01:42 AM

ABSTRACT BODY:

Abstract Body: Direct selective laser sintering (SLS) process is a novel and promising approach for additive manufacturing of various ceramic materials such as $BaTiO_3$ and ZrC. Without polymer binders or resins, a dense ceramic layer could be formed with a thickness of several hundred micrometers in which pores or microcracks are absent. The SLS process may induce ultrafast densification, metastable phase transformation, and extraordinary chemistry at grain boundaries. For example, a Ti-rich secondary phase was distributed at the grain boundaries of the SLSed hexagonal $BaTiO_3$. This talk will discuss the new phenomenon and fundamental questions related to laser-material interactions during SLS of advanced ceramics.

CONTROL ID: 3986290
CURRENT SYMPOSIUM: S2 - Advanced Microelectronics
CURRENT SESSION: Memristive switching and performance
TITLE: Computing with ions: The journey from batteries to dynamically reconfigurable adaptive architectures
PRESENTATION TYPE: Invited (by Invitation Only)
PRESENTER (FIRST NAME ONLY): A.
PRESENTER (LAST NAME ONLY): Talin
PRESENTER (INSTITUTION ONLY): Sandia National Laboratories
PRESENTER (EMAIL ONLY): aatalin@sandia.gov
PRESENTER (COUNTRY ONLY): United States
DATE/TIME SUBMITTED: September 26, 2023, 04:30 AM
ABSTRACT BODY:
Abstract Body: Acting like dopants in traditional semiconductors, electrochemically inserted ions tune the conductivity of a host by locally perturbing its electronic structure in a dynamic and reversible manner, and the resulting change in conductance can span orders of magnitude, from gradual increments needed for analog elements, to large, abrupt changes for dynamically reconfigurable adaptive architectures. In my presentation, I will discuss the recent progress in ECRAM devices spanning organic, inorganic, and 2D materials, circuits, architectures and the rich portfolio of challenging, exciting fundamental questions and how we can harness ECRAM to realize a new paradigm for low power neuromorphic computing.

CONTROL ID: 3986371

CURRENT SYMPOSIUM: S3 - Advanced Electronic Materials: Processing and Structures and Properties and Applications

CURRENT SESSION: Advanced electronic materials, including ferroelectric, piezoelectric, dielectric, electrostrictive, electrocaloric and pyroelectric materials

TITLE: New horizons for the electrocaloric cooling

PRESENTATION TYPE: Invited (by Invitation Only)

PRESENTER (FIRST NAME ONLY): Veronika

PRESENTER (LAST NAME ONLY): Kovacova

PRESENTER (INSTITUTION ONLY): Luxembourg Institute of Science and Technology

PRESENTER (EMAIL ONLY): veronika.kovacova@list.lu

PRESENTER (COUNTRY ONLY): Luxembourg

DATE/TIME SUBMITTED: September 28, 2023, 11:16 AM

ABSTRACT BODY:

Abstract Body: About 25% of the total energy produced in USA and China is spent on cooling. The main technology for cooling is well established and affordable vapour compression system. However, it uses materials emitting greenhouse gases, it is noisy and achieves moderate conversion efficiency. The electrocaloric effect-based cooling is an interesting concept for an alternative to vapour compression. It does not involve greenhouse gases emission; it is less noisy and enables 80 % of efficiency. However, a competitive demonstrator with large temperature span and cooling power is needed to convince the cooling industry.

In this work we use lead scandium tantalate to build a prototype beating a threshold for industrial applications. It consists of 140 multi-layer capacitors placed in an optimised way to enhance the temperature span. Under moderate electric field of 10 V/μm we were able to reach 20 K of temperature span between the hot and the cold end of the prototype. Moreover, we investigated its cooling power by adapting the double loop method used for elastocaloric coolers. We measured 4.2 W for a temperature span of 2.1 K. By extrapolation, the maximum cooling power was estimated to be 4.7 W (Li et al., Science, 2023, DOI : 10.1126/science.adi5477).

This study shows an electrocaloric cooler exhibiting state-of-the-art properties, manifesting that the electrocaloric cooling technology is one step closer to the cooling and air conditioning industry.

CONTROL ID: 3986466
CURRENT SYMPOSIUM: S12 - Magnetic and Superconducting Materials: From Basic Science to Applications
CURRENT SESSION: Superconducting and related magnetic materials
TITLE: Understanding phase transitions in correlated quantum materials
PRESENTATION TYPE: Invited (by Invitation Only)
PRESENTER (FIRST NAME ONLY): Panchapakesan
PRESENTER (LAST NAME ONLY): Ganesh
PRESENTER (INSTITUTION ONLY): Oak Ridge National Lab
PRESENTER (EMAIL ONLY): ganeshp@ornl.gov
PRESENTER (COUNTRY ONLY): United States
DATE/TIME SUBMITTED: September 26, 2023, 10:50 AM
ABSTRACT BODY:
Abstract Body: One of the most promising route for manipulating the properties of correlated solids for technological applications is through controlled perturbations via atomic-defects, doping, stoichiometry, strain as well as heterostructuring. However the mechanisms that drive the electronic, magnetic and/or topological transitions in these materials and the specific role of these perturbations is not fully understood. The challenge lies in predicting and understanding material responses to these perturbations from the intricate couplings of charge, spin, orbital, and lattice degrees of freedom. These at times challenge standard modeling approaches, requiring either significant empiricism or adoption of new methodologies. As such, in addition to using density functional theory, we also outline our use of the highly accurate many-body ab initio quantum Monte Carlo (QMC) approach to address these challenges. To control computational costs, we have developed a protocol of using QMC results to validate more scalable approaches via magnetic moments, charge densities, and thermodynamic properties. We present results for bulk and heterostructures of VO_2, our model for how doping controls the metal-insulator transition in the correlated-perovskites, the role of defects in inducing transitions in $MnBi_2Te_4$ and $PdCoO_2$ systems, and effects of stacking and dimensional confinement in $MnBi_2Te_4$ and $TbMn_6Sn_6$ following this protocol.

CONTROL ID: 3986507

CURRENT SYMPOSIUM: S12 - Magnetic and Superconducting Materials: From Basic Science to Applications

CURRENT SESSION: Superconducting and related magnetic materials

TITLE: Atomically thin superconducting $PdTe_2$

PRESENTATION TYPE: Invited (by Invitation Only)

PRESENTER (FIRST NAME ONLY): Wenzhuo

PRESENTER (LAST NAME ONLY): Wu

PRESENTER (INSTITUTION ONLY): Purdue University

PRESENTER (EMAIL ONLY): wenzhuowu@purdue.edu

PRESENTER (COUNTRY ONLY): United States

DATE/TIME SUBMITTED: September 26, 2023, 11:13 AM

ABSTRACT BODY:

Abstract Body: Two-dimensional (2D) van der Waals (vdW) palladium ditelluride ($PdTe_2$) has attracted tremendous interest in the past several years for studying superconductivity that approaches the 2D limit. It also hosts type-II Dirac nodes, making it a promising candidate for future exploration of the interplay between superconductivity and topological behavior. However, experimental efforts to achieve large-area atomically thin crystalline $PdTe_2$ films have been limited. Herein, we provide a novel bottom-up route to synthesize high-quality two-dimensional $PdTe_2$ films. $PdTe_2$ films with a controlled thickness ranging from a few layers to tens of nanometers and a lateral size of up to 100 µm can be obtained. The as-synthesized $PdTe_2$ films exhibited robust superconductivity at a critical temperature of 1.1 K. This study offers a new strategy for the potential monolithic integration of superconductor devices.

CONTROL ID: 3986529
CURRENT SYMPOSIUM: S14 - Emerging Semiconductor Materials and Interfaces
CURRENT SESSION: WBG and UWBG materials - synthesis, characterization, theory
TITLE: HF-Free Etching of β-Ga_2O_3
PRESENTATION TYPE: Contributed (Oral)
PRESENTER (FIRST NAME ONLY): Elizabeth
PRESENTER (LAST NAME ONLY): Sowers
PRESENTER (INSTITUTION ONLY): Air Force Research Lab
PRESENTER (EMAIL ONLY): elizabeth.sowers@us.af.mil
PRESENTER (COUNTRY ONLY): United States
DATE/TIME SUBMITTED: September 27, 2023, 07:50 AM
ABSTRACT BODY:
Abstract Body: Beta-gallium oxide (BGO) has gained attention in recent years because of its known ultra-wide bandgap, high breakdown electric field, and appropriate mobility, which makes it a promising semiconductor for high-efficiency device applications. In the fabrication process, etching is a critical step in device completion and performance. BGO is widely known to be difficult to etch, due to the strength of its ionic bonds. This difficulty limits the reagents that can etch BGO, which can lead to surface damage and low-aspect ratio architecture. A better etching technique is needed to increase device performance for this semiconductor material. We prove that BGO can be etched without plasma assistance and without the introduction of HF using microwave irradiation. Samples were characterized via atomic force microscopy (AFM) and energy dispersive X-ray spectrometry (EDS).

CONTROL ID: 3986545
CURRENT SYMPOSIUM: S10 - Fundamental mechanisms in materials for next generation computing devices
CURRENT SESSION: Phase transitions in 2D materials and devices
TITLE: Harnessing Electron Correlations and Anharmonicity for Energy Efficient Computing
PRESENTATION TYPE: Invited (by Invitation Only)
PRESENTER (FIRST NAME ONLY): Panchapakesan
PRESENTER (LAST NAME ONLY): Ganesh
PRESENTER (INSTITUTION ONLY): Oak Ridge National Lab
PRESENTER (EMAIL ONLY): ganeshp@ornl.gov
PRESENTER (COUNTRY ONLY): United States
DATE/TIME SUBMITTED: September 26, 2023, 11:40 AM
ABSTRACT BODY:
Abstract Body: Metal oxide-based Resistive Random-Access Memory (RRAM) exhibits multiple resistance states – due an underlying tunable metal-insulator transition (MIT). Similarly, ferroelectric and emerging ferro-ionic materials, particularly those with CMOS-compatible processing, are promising candidates for synaptic weight elements because of their nonvolatile multilevel memory effects. But conventional RRAM materials require high forming potentials and show high variability. Similarly, less reliability, high-power and voltage-time dilemma are suspected to plague ferroelectric synapses. To address these challenges, we are working to answer pressing questions such as: what material characteristics we need when choosing a memristor material? What factors trigger a state-change (e.g. MIT, magnetic-transition, or ferroelectric-switching) in these materials? What determines the dynamics of the switching mechanism? Using multi-scale, multi-fidelity and multi-physics theoretical methods, we discovered that harnessing electron correlations in oxides can be advantageous for improved performance of RRAM devices. For ferroelectric-based synapses, we discovered that 2D thiophosphate family of ferro-ionic materials shows strong anharmonic coupling, resulting in scale-free polarization, alleviating the voltage-time dilemma and CMOS-compatible hexagonal ferroelectrics show defect-tunable unconventional dendritic switching.

CONTROL ID: 3986558
CURRENT SYMPOSIUM: S10 - Fundamental mechanisms in materials for next generation computing devices
CURRENT SESSION: Co-design of materials, devices, and systems
TITLE: Metal-Oxide NeoHebbian Synapses for Scalable Learning
PRESENTATION TYPE: Invited (by Invitation Only)
PRESENTER (FIRST NAME ONLY): Dmitri
PRESENTER (LAST NAME ONLY): Strukov
PRESENTER (INSTITUTION ONLY): UC Santa Barbara
PRESENTER (EMAIL ONLY): strukov@ece.ucsb.edu
PRESENTER (COUNTRY ONLY): United States
DATE/TIME SUBMITTED: September 26, 2023, 11:44 AM
ABSTRACT BODY:
Abstract Body: The backpropagation through time (BPTT) algorithm is commonly used for learning and modeling temporal data. The algorithm applies the BP technique on the unrolled recurrent neural network T (~ up to 1000) times for T-step-long input sequences. Hence, training such an effective T-layer-deep network requires enormous energy and time. Recently suggested e-Prop learning algorithm addresses this inefficiency by taking inspiration from biologically plausible three-factor rules. E-prop synaptic weights are adjusted using locally stored information from pre- and post-synaptic neurons, avoiding resource-taxing network unrolling. Hence, such an algorithm is naturally suited for (in-memory, analog) neuromorphic hardware that could drastically reduce training time and energy.

In my talk, I will discuss my group's recent work on “neoHebbian” synapses essential for an efficient e-Prop algorithm hardware implementation. Such synapses feature two state variables - a neuron coupling weight and an “eligibility trace” for updating the weight. Several design options were considered to implement these features, including implementations based on single-state variable and multiple-state variable metal-oxide resistive switching devices and utilizing temperature as a state variable. I will conclude my talk by discussing promising research directions at material and device levels for improving the performance of such synapses.

CONTROL ID: 3986992
CURRENT SYMPOSIUM: S16 - New Frontiers and Science in Additive Manufacturing of Ceramic Materials
CURRENT SESSION: New applications of additively manufactured ceramics
TITLE: Micro-Printing of Ceramic Packaging for Sensor Integration
PRESENTATION TYPE: Invited (by Invitation Only)
PRESENTER (FIRST NAME ONLY): Tobias
PRESENTER (LAST NAME ONLY): Schaedler
PRESENTER (INSTITUTION ONLY): HRL Laboratories
PRESENTER (EMAIL ONLY): taschaedler@hrl.com
PRESENTER (COUNTRY ONLY): United States
DATE/TIME SUBMITTED: September 26, 2023, 06:46 PM
ABSTRACT BODY:
Abstract Body: We have developed a low-viscosity, pre-ceramic resin for recently commercialized micro-printers based on UV light projectors. This enables additive micro-manufacturing of ceramic parts on the centimeter scale with 10 μm sized features. Here we present via arrays printed directly into the ceramic with diameter and pitch as small as 10μm and 20μm, respectively. The vias are subsequently metallized by melt infiltration to provide electric pathways. This technology enables unprecedented via routing and packaging options for the 3D integration of microelectronic subsystems. We have printed interposers with thousands of vias and used single sided indium bump technology to bond them to test chips fabricated by standard semiconductor lithography. Test results including yield and resistivity will be shared. Challenges with printing curved vias and as well as integrating 3D printed parts with conventionally fabricated chips will be discussed. An outlook will be given on how this additive manufacturing technology could enable connecting curved infrared sensors with commercial-of-the-shelf (COTS) read-out integrated circuits (ROICs).

CONTROL ID: 3987292
CURRENT SYMPOSIUM: S11 - In situ/operando Characterization of Nanomaterials
CURRENT SESSION: Study of electronic materials such as Mott insulators, dielectrics, ionics, ferroics, semiconductors, superconductors etc. using advanced characterization techniques
TITLE: Phonons in ferroelectric oxides domains
PRESENTATION TYPE: Invited (by Invitation Only)
PRESENTER (FIRST NAME ONLY): Sandhya
PRESENTER (LAST NAME ONLY): Susarla
PRESENTER (INSTITUTION ONLY): Arizona State University
PRESENTER (EMAIL ONLY): sandhya.susarla@asu.edu
PRESENTER (COUNTRY ONLY): United States
DATE/TIME SUBMITTED: September 27, 2023, 01:56 AM
ABSTRACT BODY:
Abstract Body: Ferroelectric random access memory (FeRAM) that relies on polarization switching property in inorganic oxide ferroelectrics is a promising solution for low-power information storage. One of the important knobs that dictates the structure-property relationship in oxides is the phonon properties. The phonon propagation has significant impact on the emergent properties such as novel thermal and electronic conducting channels, atomic scale phononic filters, and localization and chirality in topological ferroelectrics. However, the major experimental challenge has been to probe the phonons at the "buried" thin film interfaces. In this talk, I will enlighten about vibrational electron energy loss spectroscopy, a modern day electron spectroscopy toolbox to uncover the phonons at the thin film ferroelectric substrate interfaces and the ferroelectric domains and domain walls using $PbTiO_3$ as the model system. The scientific learnings from this study could be used for designing phononic devices or ferroelectric memories that have been only predicted theoretically until now.

CONTROL ID: 3987303
CURRENT SYMPOSIUM: S14 - Emerging Semiconductor Materials and Interfaces
CURRENT SESSION: 2D materials - synthesis, characterization, theory
TITLE: Imaging moiré exctions via the low-loss electron energy loss spectroscopy
PRESENTATION TYPE: Invited (by Invitation Only)
PRESENTER (FIRST NAME ONLY): Sandhya
PRESENTER (LAST NAME ONLY): Susarla
PRESENTER (INSTITUTION ONLY): Arizona State University
PRESENTER (EMAIL ONLY): sandhya.susarla@asu.edu
PRESENTER (COUNTRY ONLY): United States
DATE/TIME SUBMITTED: September 27, 2023, 02:57 AM
ABSTRACT BODY:
Abstract Body: VanderWaal Materials when exfoliated to single layers create quantum confinement effect leading to novel electronic states. While the traditional approach has been to exfoliate bulk materials into monolayer, the modern day approach is to re-assemble different monolayers in a lego style with twist degree of freedom. The resultant structural reconstructions from aligned moiré herterostructures gives rise to novel electronic, excitonic and magnetic properties. These nanoscale interactions are often non-trivial to visualize via the traditional far-field spectroscopy technqiues. On the other hand, transmission electron microscopy coupled with electron energy loss spectroscopy (EELS) offers the advantage of sub angstrom probe size to measure electronic and excitonic states at the atomic scale. In this presentation, I will highlight about our recent progress on imaging moiré excitons: from barely detecting the moiré exciton signal in EELS to nanoscale imaging of these excitonic states scale to recognizing the fine structure of the moiré excitons. The future for probing excitons at the nanoscale under cryogenic conditions is bright with promising developments of ultralow temperature holders, electrical biasing holders, and multimodal EELS-CL technique development.

CONTROL ID: 3987384
CURRENT SYMPOSIUM: S16 - New Frontiers and Science in Additive Manufacturing of Ceramic Materials
CURRENT SESSION: New applications of additively manufactured ceramics
TITLE: Stereolithographic Additive Manufacturing of Ceramic Components with Functionally Modulated Structures
PRESENTATION TYPE: Invited (by Invitation Only)
PRESENTER (FIRST NAME ONLY): Soshu
PRESENTER (LAST NAME ONLY): Kirihara
PRESENTER (INSTITUTION ONLY): Osaka University
PRESENTER (EMAIL ONLY): kirihara@jwri.osaka-u.ac.jp
PRESENTER (COUNTRY ONLY): Japan
DATE/TIME SUBMITTED: September 27, 2023, 06:01 AM
ABSTRACT BODY:
Abstract Body: Laser scanning stereolithography of additive manufacturing techniques has been developed to create metal and ceramic components with dendritic geometries and ordered microcavities. The microlattice patterns were propagated spatially in the computer graphic space. The photosensitive resin pastes with metal and ceramic particles were photo-polymerized using an ultraviolet laser. Using a stereolithographic lamination process, solid electrolyte dendrites of yttria-stabilized zirconia (YSZ) were fabricated for fuel cell miniaturization. The gaseous fluid profiles and pressure distributions were visualized and analyzed. Subsequently, metallodielectric photonic crystals with diamond lattice structures were fabricated. The electromagnetic wave properties were measured using terahertz time-domain spectroscopy.

CONTROL ID: 3988391
CURRENT SYMPOSIUM: S3 - Advanced Electronic Materials: Processing and Structures and Properties and Applications
CURRENT SESSION: Materials design, new materials and structures, and their emerging applications
TITLE: Molecular ceramics for advanced electronic and energy applications
PRESENTATION TYPE: Invited (by Invitation Only)
PRESENTER (FIRST NAME ONLY): Julian
PRESENTER (LAST NAME ONLY): Walker
PRESENTER (INSTITUTION ONLY): Norwegian University of Science and Technoogy
PRESENTER (EMAIL ONLY): julian.walker@ntnu.no
PRESENTER (COUNTRY ONLY): Norway
DATE/TIME SUBMITTED: September 28, 2023, 12:26 PM
ABSTRACT BODY:
Abstract Body: Molecular materials were prominent during the discovery of ferroelectricity with Rochelle salt but have since dwindled. The last ten years, however, has seen a revival of interest in molecular-base polar dielectric materials, molecular ceramics. The reason being that with individual molecular components weakly bonded into a crystal lattice, these molecular ceramics exhibit intermediate functional and mechanical properties that place them in a unique parameter space between commercial metal oxides and polymers. The growing interest in these materials stems from the possibility of new functionality, low temperature synthesis and recyclability. With this work we will examine example compositions, discuss features of their electrical and electromechanical responses, approaches to compositional engineering, developments in computation and theory mechanistic understanding of their electromechanical properties and opportunities for developing their recyclability.

CONTROL ID: 3989196
CURRENT SYMPOSIUM: S5 - Controlling Grain Boundary Structure, Chemistry, and their Network as a Function of Mtl Processing
CURRENT SESSION: Nanocrystalline ceramics
TITLE: Avenues for Forming High Grain Boundary Network Ceramics with Novel Properties
PRESENTATION TYPE: Contributed (Oral)
PRESENTER (FIRST NAME ONLY): James
PRESENTER (LAST NAME ONLY): Wollmershauser
PRESENTER (INSTITUTION ONLY): U.S. Naval Research Laboratory
PRESENTER (EMAIL ONLY): james.wollmershauser@nrl.navy.mil
PRESENTER (COUNTRY ONLY): United States
DATE/TIME SUBMITTED: October 05, 2023, 02:20 PM
ABSTRACT BODY:
Abstract Body: Recent advances in processing dense nanocrystalline ceramics allows formation of ceramics with extremely small grain sizes creating solids with very high grain boundary areas which demonstrate improved properties. Coupling these processing approaches with particle atomic layer deposition (pALD) provides an opportunity to further design high grain boundary materials into nanocomposite ceramics where the properties are controlled by a coupling of the nanoscale microstructure and phase composition. This talk will cover processing approaches developed at the U.S. Naval Research Laboratory to controllably form nanocomposite ceramics where the phase constituents can all have length scales much less than 50 nanometers. The approach utilizes pALD in concert with novel and traditional ceramic sintering techniques which aim to preserve the structure, size, and phase of the initial constituents. Special considerations for processing such high surface area materials are addressed, as well as, novel properties of the formed nanocomposite ceramics.

CONTROL ID: 3989479

CURRENT SYMPOSIUM: S7 - Defects and Transport in Ceramics

CURRENT SESSION: Impact of dislocations and grain boundaries on ceramic functional properties

TITLE: Changing photoelectric and surface properties of ceramics by introducing dislocations

PRESENTATION TYPE: Contributed (Oral)

PRESENTER (FIRST NAME ONLY): Till

PRESENTER (LAST NAME ONLY): Frömling

PRESENTER (INSTITUTION ONLY): Fraunhofer IWKS

PRESENTER (EMAIL ONLY): froemling@ceramics.tu-darmstadt.de

PRESENTER (COUNTRY ONLY): Germany

DATE/TIME SUBMITTED: September 29, 2023, 02:28 AM

ABSTRACT BODY:

Abstract Body: The defect chemistry of ceramics has been extensively modified by chemical doping for decades. Nevertheless, a limit has been reached concerning this doping strategy. Therefore, we suggest to use dislocations as one-dimensional defects. These have so far been mostly disregarded as defects for modification of functional properties but are finding increasing attention recently. Given new methods of introducing dense dislocation networks in multiple ceramics, extensive investigations found a significant impact on ferroelectric and electrical properties. In this work, we elucidate the change in photoelectric and surface electrical behavior. This will rationalize whether dislocations are a new pathway for altering important parameters concerning catalysis, photocatalysis, and modification of Schottky barriers.

CONTROL ID: 3992321
CURRENT SYMPOSIUM: S14 - Emerging Semiconductor Materials and Interfaces
CURRENT SESSION: WBG and UWBG materials - synthesis, characterization, theory
TITLE: Doping the Undopable: Hybrid Molecular Beam Epitaxy Growth of Ultrawide Bandgap Oxides
PRESENTATION TYPE: Invited (by Invitation Only)
PRESENTER (FIRST NAME ONLY): Bharat
PRESENTER (LAST NAME ONLY): Jalan
PRESENTER (INSTITUTION ONLY): University of Minnesota
PRESENTER (EMAIL ONLY): bjalan@umn.edu
PRESENTER (COUNTRY ONLY): United States
DATE/TIME SUBMITTED: October 01, 2023, 06:18 PM
ABSTRACT BODY:
Abstract Body: The alkaline earth stannates are touted for their wide band gaps and the highest room-temperature electron mobilities. $CaSnO_3$ has the highest measured band gap in this family and is thus a particularly promising ultra-wide band gap semiconductor. However, discouraging results from previous theoretical studies and failed doping attempts had written off this material as "undopable". Here we redeem $CaSnO_3$ using hybrid molecular beam epitaxy (hMBE), which provides an adsorption-controlled growth for the phase-pure, epitaxial and stoichiometric $CaSnO_3$ films. By introducing lanthanum (La) as an n-type dopant, we demonstrate the robust and predictable doping of $CaSnO_3$ with free electron concentrations, n, from 3.3×10^{19} cm^{-3} to 1.6×10^{20} cm^{-3}. The films exhibit a maximum room- temperature mobility of 42 $cm^2 V^{-1} s^{-1}$ at $n = 3.3 \times 10^{19}$ cm^{-3}. Despite having a similar radius as that of the host ion, La expands the lattice parameter. Using density functional calculations, this effect is attributed to the energy gain by lowering the conduction band upon volume expansion. Finally, we exploit the robust doping by fabricating the $CaSnO_3$ -based field-effect transistors. The transistors show promise for $CaSnO_3$'s high-voltage capabilities by exhibiting low off-state leakage below 20 pA/µm at a drain-source voltage of 100 V and on-off ratios exceeding 10^6.

CONTROL ID: 3992497
CURRENT SYMPOSIUM: S2 - Advanced Microelectronics
CURRENT SESSION: Memory and data storage devices based on spintronics and ferroelectrics
TITLE: High Operating Temperature (HOT) Memory based on Ferroelectric Nitrides for 400°C Operation and Beyond
PRESENTATION TYPE: Contributed (Oral)
PRESENTER (FIRST NAME ONLY): Geoff
PRESENTER (LAST NAME ONLY): Brennecka
PRESENTER (INSTITUTION ONLY): Colorado School of Mines
PRESENTER (EMAIL ONLY): gbrennec@mines.edu
PRESENTER (COUNTRY ONLY): United States
DATE/TIME SUBMITTED: October 04, 2023, 10:26 AM
ABSTRACT BODY:
Abstract Body: Wurtzite nitride ferroelectrics such as (Al,Sc)N and (Al,B)N offer a number of potential advantages and challenges— in particular, the large spontaneous polarization and coercive field values when compared to other ferroelectrics—for device integration. These nitride ferroelectrics are particularly attractive for applications in High Operating Temperature (HOT) applications, such as on Venus, in geothermal exploration, and other warm environments. We report here the growth, integration, and ferroelectric switching behavior of (Al,Sc)N and (Al,B)N films with SiC-based electronics as important steps towards full HOT memory devices. Increased operating temperatures lead to significant reductions in coercive fields but have very little impact on switchable polarization. At elevated temperatures, switching endurance degrades and is voltage-dependent, leading to a situation where imprint may impact retention performance.

CONTROL ID: 3993404

CURRENT SYMPOSIUM: S3 - Advanced Electronic Materials: Processing and Structures and Properties and Applications

CURRENT SESSION: Performance, reliability, and fatigue of ferroelectrics and related devices

TITLE: Importance of Cation Redox in the Pressureless Sintering and Subsequent Properties of Entropy-Stabilized J14 Ceramics

PRESENTATION TYPE: Invited (by Invitation Only)

PRESENTER (FIRST NAME ONLY): Geoff

PRESENTER (LAST NAME ONLY): Brennecka

PRESENTER (INSTITUTION ONLY): Colorado School of Mines

PRESENTER (EMAIL ONLY): gbrennec@mines.edu

PRESENTER (COUNTRY ONLY): United States

DATE/TIME SUBMITTED: October 04, 2023, 10:28 AM

ABSTRACT BODY:

Abstract Body: Since Rost's first report of entropy stabilization of rocksalt-phase (Mg,Co,Ni,Cu,Zn)O (aka J14), interest in both this particular chemistry and the general concept of entropy stabilization has exploded. The majority of reports, however, focus solely on properties, with very little attention paid to fundamental sintering and densification. Most studies on J14 ceramics, for example, are based around either low-density (e.g., <80%) samples or resort to spark plasma sintering or other similar techniques. Focusing on cation diffusion during sintering, we investigate the precipitation of Cu species under low-pO_2 conditions and have found that the oxidation and reduction of $Co^{2+} \leftrightarrow Co^{3+}$ plays a crucial role in the diffusion kinetics and densification of J14 ceramics when sintering in air. Accommodating this redox reaction during sintering enables simple and reliable pressureless sintering of J14 ceramics to ~98% relative density. Naturally, defect chemistry and compensation also dominate the charge transport in dense samples. This work details a simple approach to densify J14 but also serves as a reminder of the importance of defect chemistry to sintering and densification in ceramics of many complex compositons.

CONTROL ID: 3996633
CURRENT SYMPOSIUM: S7 - Defects and Transport in Ceramics
CURRENT SESSION: Point defect segregation to or depletion from dislocations, surfaces, grain boundaries, and interfaces
TITLE: Three-dimensional nanoscale measurement of oxygen in oxides using atom probe tomography
PRESENTATION TYPE: Contributed (Oral)
PRESENTER (FIRST NAME ONLY): David
PRESENTER (LAST NAME ONLY): Diercks
PRESENTER (INSTITUTION ONLY): Colorado School of Mines
PRESENTER (EMAIL ONLY): ddiercks@mines.edu
PRESENTER (COUNTRY ONLY): United States
DATE/TIME SUBMITTED: October 05, 2023, 02:54 PM
ABSTRACT BODY:
Abstract Body: The ionic transport properties in oxides used for next generation fuel cells, batteries, and other solid state energy devices are dependent, and largely limited, by what happens at grain boundaries and phase boundaries. Therefore, the understanding and quantification of species in these locations provides important insight into nanoscale transport and what approaches can be taken to improve it. However, the tools which can directly measure such phenomena at the nanoscale, particularly quantitatively and in three-dimensions have been lacking. Here, atom probe tomography is applied to exploring oxygen and cation concentrations in relation to grain and phase boundaries. Examples will include work on cerias, chromites, manganites, and zirconates. In some of these cases oxygen-isotope exchange was used to introduce ^{18}O into the structure after fabrication to allow for differentiation between the oxygen in the original structure and that which subsequently diffused into the material. Using the nanoscale compositional information from atom probe tomography along with finite element modeling, mass transport parameters such as grain boundary oxygen diffusivity and the surface exchange coefficient could be extracted. Results also demonstrate boundary-to-boundary variations suggestive of the importance of the local crystallography as well.

CONTROL ID: 3997873
CURRENT SYMPOSIUM: S15 - Electronics Materials for 5G/6G Applications
CURRENT SESSION: Materials growth and design
TITLE: Glass as a Dielectric Material for 5G Interconnect and Quantum Transducer Packages
PRESENTATION TYPE: Invited (by Invitation Only)
PRESENTER (FIRST NAME ONLY): Michael
PRESENTER (LAST NAME ONLY): Lanagan
PRESENTER (INSTITUTION ONLY): Penn State University
PRESENTER (EMAIL ONLY): mxl46@psu.edu
PRESENTER (COUNTRY ONLY): United States
DATE/TIME SUBMITTED: October 03, 2023, 01:15 PM
ABSTRACT BODY:
Abstract Body: Glass substrates with a variety of compositions are commercially available, from high purity silica with ppm impurity levels to surface strengthened glass with substantial amounts of alkali and alkaline earth modifiers. The dielectric and optical properties were characterized over a broad frequency range between 1 MHz and 18 THz. Fundamental dielectric relaxation mechanisms in the 5G to 6G frequency range will be described for amorphous and crystalline materials along with their impact on dielectric permittivity, loss and temperature dependence of permittivity. Glass substrates are compatible with semiconductor processes where surfaces with sub-nanometer roughness are ideal for creating transmission line structures for 5G packages and direct bonding of multiple substrates for hermetically sealed optical quantum transducers. The relative contributions of dielectrics and conductors to transmission line loss in 5G device performance will be discussed.

CONTROL ID: 3998105
CURRENT SYMPOSIUM: S12 - Magnetic and Superconducting Materials: From Basic Science to Applications
CURRENT SESSION: Other magnetic materials
TITLE: Tailoring Physical Properties of 2D Magnets through Synthetic Parameters
PRESENTATION TYPE: Contributed (Oral)
PRESENTER (FIRST NAME ONLY): Wenhao
PRESENTER (LAST NAME ONLY): Liu
PRESENTER (INSTITUTION ONLY): University of Texas, Dallas
PRESENTER (EMAIL ONLY): wenhao.liu@utdallas.edu
PRESENTER (COUNTRY ONLY): United States
DATE/TIME SUBMITTED: October 03, 2023, 05:33 PM
ABSTRACT BODY:
Abstract Body: Growth parameters play a significant role in the crystal quality and physical properties of layered materials. Here we present a case study on a van der Waals (vdW) magnetic $NbFeTe_2$ material. Two different types of polymorphic $NbFeTe_2$ phases, as large as 5x5 mm synthesized at different temperatures, display significantly different behaviors in crystal symmetry, electronic structure, electrical transport, and magnetic properties. While the phase synthesized at low temperature showing behavior consistent with previous reports, the new phase synthesized at high temperature, has completely different physical properties, such as metallic resistivity, long-range ferromagnetic order, anomalous Hall effect, negative magnetoresistance, and distinct electronic structures. Besides the $NbFeTe_2$, a series of vdW magnets such as $NbCoTe_2$, $NbNiTe_2$, $TaCoTe_2$ and $TaNiTe_2$ crystals have been synthesized and the effect of synthesis condition on their properties have been thoroughly investigated. Our work demonstrates that simply tuning synthetic parameters in a controlled manner could be an effective route to alter the physical properties of existing materials, potentially unlock new states of matter, or even discover new materials.

CONTROL ID: 3998364
CURRENT SYMPOSIUM: S15 - Electronics Materials for 5G/6G Applications
CURRENT SESSION: 5G/6G Metrology
TITLE: Polar Vortices films at mmWave frequencies
PRESENTATION TYPE: Contributed (Oral)
PRESENTER (FIRST NAME ONLY): Florian
PRESENTER (LAST NAME ONLY): Bergmann
PRESENTER (INSTITUTION ONLY): NIST
PRESENTER (EMAIL ONLY): florian.bergmann@nist.gov
PRESENTER (COUNTRY ONLY): United States
DATE/TIME SUBMITTED: October 04, 2023, 10:56 AM
ABSTRACT BODY:
Abstract Body: $PbTiO_3$-$SrTiO_3$ superlattice films form Polar Vortices under the right lattice periodicity conditions. Experiments at kHz and THz frequencies have already shown emergent properties of these superlattice films like negative permittivity and collective lattice dynamics. However, data in the GHz range are sparse. Measurements in this frequency range are both interesting for the fundamental understanding of Polar Vortices and for potential electronic applications of the films. Here, we report on broadband measurements from 70 kHz to 110 GHz using Coplanar Waveguide devices. We extract the permittivity of a PbTiO3-SrTiO3 superlattice film on a $DyScO_3$ substrate and compare the results to those from co-fabricated $PbTiO_3$ and $SrTiO_3$ films. For all samples, we characterize the complex in-plane permittivity in two perpendicular orientations and as a function of bias voltage. Based on our results, we discuss the potential of Polar Vortex films for mmWave applications.

CONTROL ID: 3998384
CURRENT SYMPOSIUM: S4 - Complex Oxide Thin Films and Heterostructures: From Synthesis to Strain/Interface-engineered Emergent Properties
CURRENT SESSION: Characterizations of strain, defects, and interfaces
TITLE: Spatially Resolved Mosaicity in Wurtzite-Structured (Al,Sc)N Thin Films via DPC STEM
PRESENTATION TYPE: Contributed (Oral)
PRESENTER (FIRST NAME ONLY): Charles
PRESENTER (LAST NAME ONLY): Evans
PRESENTER (INSTITUTION ONLY): Carnegie Mellon University
PRESENTER (EMAIL ONLY): charlese@andrew.cmu.edu
PRESENTER (COUNTRY ONLY): United States
DATE/TIME SUBMITTED: October 09, 2023, 09:09 PM
ABSTRACT BODY:
Abstract Body: Aluminum-scandium nitride – (Al,Sc)N – is a polar, wurtzite-structured material which has recently been shown to be ferroelectrically switchable. While the switching mechanisms are not yet fully understood, one factor which may affect the ferroelectric properties of these films is their grain structure, such as the degree of mosaicity, which can be affected by factors such as the choice of growth substrate. In addition, prior work has shown that the properties of these films may change with thickness, so it is also interesting to explore how the degree of mosaicity in these films changes through thickness.

While there are a variety of substrates which can support the growth of (Al,Sc)N films, this work is focused on (Al,Sc)N grown on gallium nitride – GaN – and indium-gallium nitride – (In,Ga)N – with a composition selected for a lattice mismatch of < 1%. The structures of the films have been imaged using differential phase contrast (DPC) scanning transmission electron microscopy (STEM). In this work it has been observed that the first few tens of nanometers of these films have a higher degree of mosaicity than the more developed parts of the film. Grains tend to become larger in-plane as film thickness increases. Additionally, atomic resolution integrated DPC (iDPC) imaging has been used to assess the atomic scale structure of these films near the interface.

CONTROL ID: 3998556

CURRENT SYMPOSIUM: S3 - Advanced Electronic Materials: Processing and Structures and Properties and Applications

CURRENT SESSION: Advanced electronic materials, including ferroelectric, piezoelectric, dielectric, electrostrictive, electrocaloric and pyroelectric materials

TITLE: Homovalent- and heterovalent-substituted $BaTiO_3$: Unravelling the origin of relaxor behavior

PRESENTATION TYPE: Invited (by Invitation Only)

PRESENTER (FIRST NAME ONLY): Giovanna

PRESENTER (LAST NAME ONLY): Canu

PRESENTER (INSTITUTION ONLY): National Research Council of Italy

PRESENTER (EMAIL ONLY): giovanna.canu@cnr.it

PRESENTER (COUNTRY ONLY): Italy

DATE/TIME SUBMITTED: October 09, 2023, 03:19 PM

ABSTRACT BODY:

Abstract Body: Barium titanate ($BaTiO_3$, BT) is not only the base material for billions of capacitors per year: possibilities for new applications are often sought after for BT.

The application of BT as a dielectric material relies on the incorporation of suitable foreign ions. Increasing compositional disorder at the atomic scale leads to a transition from the ferroelectric state to relaxor behavior, relevant for microwave, electrocaloric, and energy storage applications. However, the relationship between the macroscopic properties useful for applications and the local structure is not fully understood yet, and the origin of relaxor behavior remains unclear, with aspects dependent on the specific system.

We present here the results of an investigation on homovalent- and heterovalent-substituted BT on the titanium site, $Ba(M,Ti)O_3$ (M=Zr, Ce, Nb), over a broad range of compositions, combining macroscopic property measurements with techniques sensitive to the local structure and ab initio modeling. Zr^{4+} and Ce^{4+} are homovalent substituents for BT, with Ce^{4+} having the largest ionic radius, thus making $Ba(Ce,Ti)O_3$ a model system to study the effect of local strain; Nb^{5+} is instead a heterovalent substituent, thus $Ba(Nb,Ti)O_3$ is interesting to study local charge effects. By comparing our results with the literature, some general conclusions are drawn on the origin of relaxor behavior in BT-based systems.

CONTROL ID: 3998724
CURRENT SYMPOSIUM: S4 - Complex Oxide Thin Films and Heterostructures: From Synthesis to Strain/Interface-engineered Emergent Properties
CURRENT SESSION: Controlled synthesis of epitaxial thin films and nanocomposites, including heterostructures, superlattices, and vertically aligned nanocomposites for functional properties
TITLE: Resolving the Isopolar Interface of $LaNiO_3/LaFeO_3$ Superlattices
PRESENTATION TYPE: Contributed (Oral)
PRESENTER (FIRST NAME ONLY): Le
PRESENTER (LAST NAME ONLY): Wang
PRESENTER (INSTITUTION ONLY): Pacific Northwest National Laboratory
PRESENTER (EMAIL ONLY): le.wang@pnnl.gov
PRESENTER (COUNTRY ONLY): United States
DATE/TIME SUBMITTED: October 09, 2023, 06:12 PM
ABSTRACT BODY:
Abstract Body: Interfacial charge transfer holds the potential to generate emergent electronic and magnetic phenomena within complex oxides. The prediction of high T_c cuprate-like superconductivity physics in $LaNiO_3/LaAlO_3$ superlattices has motivated extensive research efforts in investigating various heterostructures and superlattices involving nickelates. However, to date very little is known about the isopolar interface between the antiferromagnetic charge-transfer insulator $LaFeO_3$ and the paramagnetic $LaNiO_3$. Here, we have synthesized a series of $LaNiO_3/LaFeO_3$ superlattices and investigated the charge distributions by employing in situ X-ray photoemission spectroscopy, X-ray absorption spectroscopy, and four-dimensional scanning transmission electron microscopy combined with electron energy-loss spectroscopy. We demonstrate that the charge transfer from Fe to Ni sites occurs near the interface, which strongly affects the in-plane transport behavior of the superlattices. Furthermore, density functional theory calculations confirm that the Fe^{4+}-Ni^{2+} configuration at the $LaNiO_3/LaFeO_3$ interface is more stable than the Fe^{3+}-Ni^{3+} configuration, consistent with our experimental findings of charge transfer from Fe to Ni. Overall, our results provide insights into comprehending the impact of interfacial charge transfer on the design of innovative functional oxide heterostructures with emerging properties.

CONTROL ID: 3998793
CURRENT SYMPOSIUM: S12 - Magnetic and Superconducting Materials: From Basic Science to Applications
CURRENT SESSION: Other magnetic materials
TITLE: Multiferroic-based Magnetic Control Valve for Switchless Direct Antenna Modulation
PRESENTATION TYPE: Invited (by Invitation Only)
PRESENTER (FIRST NAME ONLY): Greg
PRESENTER (LAST NAME ONLY): Carman
PRESENTER (INSTITUTION ONLY): UCLA
PRESENTER (EMAIL ONLY): carman@seas.ucla.edu
PRESENTER (COUNTRY ONLY): United States
DATE/TIME SUBMITTED: October 05, 2023, 09:44 AM
ABSTRACT BODY:
Abstract Body: Direct Antenna Modulation (DAM) techniques are used to produce power-efficient high bandwidth electrically small transmitting antenna. For magnetic based antennas, radiation frequency modulation can be accomplished through changing the antenna's biasing magnetic field to both shift the radiation frequency as well as transfer energy between the antenna's resonating states. A magnetic control valve (MCV) made of a multiferroic stack with piezoelectric and magnetoelastic materials and a permanent magnet has been developed to produce a dynamic magnetic biasing field. A dynamic voltage applied to the piezoelectric produces stress modulations in the magnetoelastic material, which in turn modulates the magnetization M creating a local oscillating magnetic fields ΔH. In this presentation, we present the MCV's operating principles constructed from Galfenol or Terfenol-D mechanically coupled with a piezoelectric material and biased with a permanent magnet. The Galfenol MCV produces ΔH~ 15 Oe, depending upon permanent magnetic, change which is attributed to magnetic pinning features introduced during the fabrication process. However, Terfenol-D experimental results show a ΔH= 50 Oe in both quasi-static and dynamic tests up to 1.5 kHz frequencies. The field produced by the Terfenol-D MCV exceeds the required field to achieve DAM.

CONTROL ID: 3998809

CURRENT SYMPOSIUM: S5 - Controlling Grain Boundary Structure, Chemistry, and their Network as a Function of Mtl Processing

CURRENT SESSION: Defect and segregation behavior

TITLE: Examining the effect of charge and size mismatch on the segregation behavior of dopants in $BaZrO_3$

PRESENTATION TYPE: Contributed (Oral)

PRESENTER (FIRST NAME ONLY): Julian

PRESENTER (LAST NAME ONLY): Ebert

PRESENTER (INSTITUTION ONLY): Forschungszentrum Juelich

PRESENTER (EMAIL ONLY): ju.ebert@fz-juelich.de

PRESENTER (COUNTRY ONLY): Germany

DATE/TIME SUBMITTED: October 09, 2023, 02:33 PM

ABSTRACT BODY:

Abstract Body: Y-doped $BaZrO_3$ is one of the materials heavily researched for intermediate-temperature proton-conducting SOFCs. Grain boundaries are found to lower the proton conductivity of this material. This can be understood according to the space charge theory by a positive core charge of the grain boundary resulting in a barrier for proton transport across the grain boundary. This study analyzes the influence of B-site doping simultaneously with two dopants of varying charge or size mismatch on the segregation behavior to the grain boundary. Y and Sc, Y and Al, and Y and Ti were chosen as the dopant combinations for this study besides a pure Y doping. The segregation behavior of the different dopants at grain boundaries was then analyzed with S/TEM and EDS, while the ionic conductivity was measured by EIS. By using ions with different ionic sizes and valencies, it is possible to evaluate the impact of these factors on segregation. This information can then be used to engineer the grain boundary segregation for higher conductivity.

CONTROL ID: 3999211
CURRENT SYMPOSIUM: S10 - Fundamental mechanisms in materials for next generation computing devices
CURRENT SESSION: CMOS compatible ferroelectric materials and devices
TITLE: Characterization of Ferroelectric Mg-Alloyed Zinc Oxide by Transmission Electron Microscopy
PRESENTATION TYPE: Poster
PRESENTER (FIRST NAME ONLY): Ece
PRESENTER (LAST NAME ONLY): Gunay
PRESENTER (INSTITUTION ONLY): Carnegie Mellon University
PRESENTER (EMAIL ONLY): egunay@andrew.cmu.edu
PRESENTER (COUNTRY ONLY): United States
DATE/TIME SUBMITTED: October 09, 2023, 10:30 PM
ABSTRACT BODY:
Abstract Body: In the pursuit of advanced electronic materials, wurtzite $Zn_{1-x}Mg_xO$ has garnered attention for its promising ferroelectric properties, offering opportunities for innovations in non-volatile memory, integrated circuits, piezoelectric devices, and energy harvesting applications. This study focuses on the characterization of ferroelectric $Zn_{1-x}Mg_xO$ thin films grown via sputter deposition. We employ transmission electron microscopy techniques, including electron diffraction, bright-field and dark-field imaging, high angle annular dark field STEM and energy-dispersive X-ray spectroscopy, to investigate the microstructure and local chemical composition of $Zn_{1-x}Mg_xO$ thin films over a range of compositions. The films grow along the c-axis of the wurtzite structure and are chemically homogeneous at least up to 35% Mg, even though this is well beyond the bulk solubility limit. The electron microscopy analyses provide valuable insights into the nanoscale structural and chemical properties of $Zn_{1-x}Mg_xO$ thin films and heterostructures, which is essential to understanding polarization switching in this novel class of ferroelectric materials.

Acknowledgments
This material is based upon work supported by the Center for 3D Ferroelectric Microelectronics (3DFeM), an Energy Frontier Research Center funded by the U.S. Department of Energy, Office of Science under Award Number DE-SC0021118.

CONTROL ID: 3999239
CURRENT SYMPOSIUM: S14 - Emerging Semiconductor Materials and Interfaces
CURRENT SESSION: WBG and UWBG materials - synthesis, characterization, theory
TITLE: A Study of the Phase Stabilization of α-Ga_2O_3 Grown on R-plane Sapphire via MOCVD
PRESENTATION TYPE: Contributed (Oral)
PRESENTER (FIRST NAME ONLY): Cheng-Han
PRESENTER (LAST NAME ONLY): Lee
PRESENTER (INSTITUTION ONLY): Cornell University
PRESENTER (EMAIL ONLY): cl2376@cornell.edu
PRESENTER (COUNTRY ONLY): United States
DATE/TIME SUBMITTED: October 09, 2023, 09:19 PM
ABSTRACT BODY:
Abstract Body: The ultra-wide bandgap semiconductor material α-Ga_2O_3, with a bandgap of 5.3 eV, enables a high breakdown voltage and a potentially high Baliga's figure of merit, thus exhibiting promising potential for applications in next-generation high-power electronics. α-Ga_2O_3, however, is a metastable polymorph of Ga_2O_3 and has a strong tendency to revert back to the thermodynamically stable β-phase. Synthesizing thick, phase-pure α-Ga_2O_3 epitaxial film with high crystallinity via MOCVD has proven to be a major challenge.

In this study, we explored the MOCVD growth window for stabilizing α-Ga_2O_3 thin films on r-plane sapphire substrate. A two-step growth procedure was employed to suppress defect generation where a 10-nm-thick nucleation layer was deposited at a low growth rate ~0.4 μm/hr prior to the growth of the thick epilayer. We identified an optimal α-Ga_2O_3 growth regime characterized by a low growth temperature 575 °C and a high growth rate > 0.85 μm/hr for the Ga_2O_3 epilayer. By employing the highlighted optimum growth conditions, we have demonstrated the stabilization of phase-pure α-Ga_2O_3 thin films with a thickness of 630 nm on r-plane sapphire, resulting in one of the highest α-Ga_2O_3 critical thickness ever reported on r-plane α-Al_2O_3 via MOCVD and an enhanced crystalline quality characterized by the low rocking curve FWHM ~1145 arcsec based on the inclusion of a low-growth-rate nucleation layer.

CONTROL ID: 3999462
CURRENT SYMPOSIUM: S17 - Electronic and Ionic Materials in Battery Energy Storage
CURRENT SESSION: Electrodes (Anodes and Cathodes)
TITLE: The Decoupling of Particle Size with Lithium Diffusion & Reaction Times in Polycrystalline Battery Particles
PRESENTATION TYPE: Contributed (Oral)
PRESENTER (FIRST NAME ONLY): Yiyang
PRESENTER (LAST NAME ONLY): Li
PRESENTER (INSTITUTION ONLY): University of Michigan
PRESENTER (EMAIL ONLY): yiyangli@umich.edu
PRESENTER (COUNTRY ONLY): United States
DATE/TIME SUBMITTED: October 06, 2023, 12:11 PM
ABSTRACT BODY:
Abstract Body: Polycrystalline secondary particles composed of Li(Ni,Mn,Co)O2 (NMC) are the most commonly used cathode materials for Li-ion batteries. During electrochemical lithiation, it is believed that lithium enters the secondary particle at the surface, and diffuses through the bulk and grain boundaries. Based on this picture, it is believed that smaller particles would lithiate faster, due to shorter diffusion lengths and larger surface-area-to-volume ratios. In this work, we develop a novel, high-throughput single-particle electrochemistry platform to evaluate this widespread assumption. By measuring the reaction and diffusion times for 21 individual particles, we find no correlation between the particle size with the reaction and diffusion time, in stark contrast to the prevailing understanding of ion insertion. We propose that electrochemical reactions occur not only at the surface, but in the bulk of the secondary particle, possibly as a result of electrolyte penetration into particle cracks. Our high-throughput, single-particle electrochemical platform opens a new frontier to enable robust, statistical quantification of individual particles in batteries and other electrochemical systems.

J. Min et al. “Direct measurements of size-independent lithium diffusion and reaction times in individual polycrystalline battery particles.” Energy & Environmental Science, 16, 3847, 2023

CONTROL ID: 3999477
CURRENT SYMPOSIUM: S7 - Defects and Transport in Ceramics
CURRENT SESSION: Defect-mediated properties
TITLE: Defects in $Zn_{1-x}Mg_xO$ Films and Their Influence on Ferroelectric Properties for Energy-Efficient Non-Volatile Memory Applications
PRESENTATION TYPE: Contributed (Oral)
PRESENTER (FIRST NAME ONLY): Pedram
PRESENTER (LAST NAME ONLY): Yousefian
PRESENTER (INSTITUTION ONLY): Pennsylvania State University
PRESENTER (EMAIL ONLY): pky5045@psu.edu
PRESENTER (COUNTRY ONLY): United States
DATE/TIME SUBMITTED: October 09, 2023, 09:29 PM
ABSTRACT BODY:
Abstract Body: As the global demand to minimize CO_2 emissions grows, energy-efficient solutions are critical in the realm of computing, which currently accounts for 5-15% of global energy consumption. $Zn_{1-x}Mg_xO$ (ZMO), a recently discovered ferroelectric oxide with the wurtzite structure, emerges as a promising candidate for advancing energy-efficient non-volatile memory. Notably, ZMO offers compelling advantages, including the ability to be grown at low processing temperatures (<400 °C) large polarizations, and excellent data retention. Nonetheless, a significant knowledge gap exists concerning defects within ZMO films and their influence on ferroelectric properties. This research explores the conduction mechanisms in ZMO films. It was found, by preparing different top electrodes that modified Poole-Frenkel emission dominates charge transport. Additionally, the defects in ZMO films were investigated using modulus spectroscopy and charge deep level transient spectroscopy (Q-DLTS). It was found that in some cases, heat treatments on metastable ZMO films induce segregation of a second phase with the rocksalt structure. This reduces remanent polarization and induces higher conductivity regions within ZMO films.

CONTROL ID: 3999486

CURRENT SYMPOSIUM: S18 - Structure and function in high-entropy, entropy-stabilized, and related oxides

CURRENT SESSION: Understanding and predicting structure and properties through computation and data science

TITLE: Oxygen vacancy formation energetics in MgO-based high entropy oxides from DFT and experimental validation

PRESENTATION TYPE: Contributed (Oral)

PRESENTER (FIRST NAME ONLY): Dilpuneet

PRESENTER (LAST NAME ONLY): Aidhy

PRESENTER (INSTITUTION ONLY): Clemson University

PRESENTER (EMAIL ONLY): daidhy@clemson.edu

PRESENTER (COUNTRY ONLY): United States

DATE/TIME SUBMITTED: October 06, 2023, 12:58 PM

ABSTRACT BODY:

Abstract Body: In contrast to single or double cation based oxides, high entropy oxides (HEOs) consist of equi- or near equimolar concentation of multiple cations randomly distributed on a crystal lattice. The random distribution causes a wide diversity of nearest-neigbor environments around oxygen atoms. Consequently, a range of bonding environments and oxygen vacancy formation energies are observed. We investigate oxygen vacancy formation energies in Mg(ZnCoNiCu)O-based HEO using density functional theory (DFT). We find that vacancy formation energy trends are controlled by the cation charge. We explicitly determine the effect of all cations on formation energies. Experimentally, a series of binary, ternary, quaternary and quinary HEOs are synthesized using a simple and scalable polymeric steric entrapment method. By annealing in inert environment, oxygen vacancies are analyzed by XPS and TGA. The experimental obseravtions agree with our DFT calcilations. This understanding contributes to oxygen vacancy transport useful in microstructure evolution and electrochemical applications.

CONTROL ID: 3999874
CURRENT SYMPOSIUM: S2 - Advanced Microelectronics
CURRENT SESSION: Flexible electronics
TITLE: Flexible Nanocomposite Materials for Energy Harvesting and Heat Management
PRESENTATION TYPE: Contributed (Oral)
PRESENTER (FIRST NAME ONLY): Zdravko
PRESENTER (LAST NAME ONLY): Kutnjak
PRESENTER (INSTITUTION ONLY): Jozef Stefan Institute
PRESENTER (EMAIL ONLY): zdravko.kutnjak@ijs.si
PRESENTER (COUNTRY ONLY): Slovenia
DATE/TIME SUBMITTED: October 07, 2023, 11:10 AM
ABSTRACT BODY:
Abstract Body: Today's quest for sustainable energy solutions through greener energy harvesting and heat-management technologies has recently developed a significant interest in new flexible and biocompatible nanocomposite ceramics with large electromechanical, triboelectric, and electrocaloric (EC) effects [1]. Therefore, an overview of experimental and theoretical investigations of the large EC, piezoelectric, and triboelectric response in flexible ceramic nanocomposites exploiting enhanced multiferroic properties of ferroelectric nanoparticles within the polymer matrix will be presented in this contribution. Specifically, the enhanced EC response in lead-free BCZT and BaTiO3-based nanoparticles will be reviewed, including flexible polymer composites' large energy harvesting potential [2]. The impact of filler's dielectric permittivity and aspect ratio in high-k polymer and the benefits of combining 1D and 3D nanofillers on enhanced properties of flexible nanocomposites will be discussed [3]. [1] Z. Kutnjak., B. Rozic, R. Pirc., Wiley Encyclopedia of Electrical and Electronics Engineering, 1-19 (2015). [2] Z. Hanani et al., Nano Energy 81, 105661 (2021). [3] Z. Hanani et al., Nanoscale Adv. 4, 4658 (2022).

CONTROL ID: 3999877

CURRENT SYMPOSIUM: S3 - Advanced Electronic Materials: Processing and Structures and Properties and Applications

CURRENT SESSION: Advanced electronic materials, including ferroelectric, piezoelectric, dielectric, electrostrictive, electrocaloric and pyroelectric materials

TITLE: Lead-free multiferroic ceramic materials with enhanced ferroelectric and electrocaloric properties

PRESENTATION TYPE: Contributed (Oral)

PRESENTER (FIRST NAME ONLY): Brigita

PRESENTER (LAST NAME ONLY): Rozic

PRESENTER (INSTITUTION ONLY): Jozef Stefan Institute

PRESENTER (EMAIL ONLY): brigita.rozic@ijs.si

PRESENTER (COUNTRY ONLY): Slovenia

DATE/TIME SUBMITTED: October 07, 2023, 11:12 AM

ABSTRACT BODY:

Abstract Body: The request for greener heat-management technologies has recently developed a significant interest in new electrocaloric (EC) effect-based cooling devices that can replace the existing cooling technics [1-2]. An overview of experimental and theoretical investigations of the large EC and piezoelectric response in lead-free ceramic composites and ceramics near antiferroelectric transition will be presented in this contribution [3-4]. Specifically, the large EC response observed by direct experiments in lead-free BCTZ-based ceramics will be reviewed, including polymer composites' considerable energy harvesting potential [5]. Besides, it is demonstrated that both negative and positive EC responses can be arbitrarily invoked in antiferroelectric materials by adequately controlling the electric field and temperature, which enables enhancement of the electrocaloric cooling power for up to 100%. [1] A. S. Mischenko et al., Science 311, 1270 (2006). [2] Z. Kutnjak., B. Rozic, R. Pirc., Wiley Encyclopedia of Electrical and Electronics Engineering, 1-19 (2015). [3] A. Bradesko et al., Acta Materialia 169, 275 (2019). [4] U. Plaznik et al., International journal of refrigeration 98, 139 (2019). [5] Z. Hanani et al., Nano Energy 81, 105661 (2021).

CONTROL ID: 3999996

CURRENT SYMPOSIUM: S16 - New Frontiers and Science in Additive Manufacturing of Ceramic Materials

CURRENT SESSION: New approaches for AM processes of ceramics and ceramic-matrix composites

TITLE: From Flash Sintering to Ultrafast Sintering without an Applied Electric Field: Review of Mechanistic Studies and Discussion of Potential Relevance to Additive Manufacturing

PRESENTATION TYPE: Invited (by Invitation Only)

PRESENTER (FIRST NAME ONLY): Jian

PRESENTER (LAST NAME ONLY): Luo

PRESENTER (INSTITUTION ONLY): University of California, San Diego

PRESENTER (EMAIL ONLY): jluo@ucsd.edu

PRESENTER (COUNTRY ONLY): United States

DATE/TIME SUBMITTED: October 08, 2023, 04:02 AM

ABSTRACT BODY:

Abstract Body: This talk will first review a series of our mechanistic studies of flash sintering [Viewpoint: Scripta 146: 260 (2018)]. First, we showed that flash sintering generally starts as a thermal runaway and developed a quantitative model [Acta 94: 87 (2015)]. Second, we showed that ultrahigh heating rates on the order of 10^2 K/s enable ultrafast sintering with and without applied electric fields in flash sintering (with Joule heating) vs. rapid thermal annealing (with intense IR heating) [Acta 125: 465 (2017)]. Subsequently, a general ultrafast high-temperature sintering method was developed [Science 368: 521 (2020)]. Two possible underlying mechanisms will be discussed. In addition, potential applications and relevance of the ultrafast sintering mechanisms, enabled by ultrahigh heating rates on the order of 10^2 K/s and a short duration (typically seconds) at a high temperature, to additive manufacturing will be discussed. Ultrafast high-temperature sintering has been applied to sinter small 3D printed artifacts [see, e.g., Science 368: 521 (2020)]. Selective laser sintering also shares similar character of ultrafast high-temperature sintering; thus, similar ultrafast sintering mechanisms may apply (albeit different technical challenges exist).

CONTROL ID: 4000041

CURRENT SYMPOSIUM: S1: Characterization of Structure-Property Relationships in Functional Ceramics

CURRENT SESSION: Advances in connecting local and global structure to properties

TITLE: Polarization and Switching in Novel Ferroelectrics: Atomic -scale Analysis via Scanning Transmission Electron Microscopy

PRESENTATION TYPE: Invited (by Invitation Only)

PRESENTER (FIRST NAME ONLY): Elizabeth

PRESENTER (LAST NAME ONLY): Dickey

PRESENTER (INSTITUTION ONLY): Carnegie Mellon University

PRESENTER (EMAIL ONLY): ecdickey@cmu.edu

PRESENTER (COUNTRY ONLY): United States

DATE/TIME SUBMITTED: October 08, 2023, 10:06 AM

ABSTRACT BODY:

Abstract Body: The recent development of novel ferroelectrics (e.g. based on HfO_2, AlN and ZnO) with robust polarization reversal and CMOS-compatible processing has provided new opportunities for integrated ferroelectrics. To understand the mechanisms that control and induce switchable states in such materials, we use scanning transmission electron microscopy (STEM) differentiated differential phase contrast (dDPC) imaging to locate projected atomic column positions to calculate partial pair distribution functions and local spontaneous polarization in thin-film ferreoelectrics. The experimental data are complemented with STEM image simulations to determine the influence of the sample and experimental variables on the structural analysis. This talk will describe the local- to meso-scale structure structural and chemical insights we have gleaned through STEM-based microscopy to understand polarization phenomena in these novel materials.

Acknowledgements

This material is based upon work supported by the Center for 3D Ferroelectric Microelectronics (3DFeM), an Energy Frontier Research Center funded by the U.S. Department of Energy, Office of Science under Award Number DE-SC0021118.

CONTROL ID: 4000069
CURRENT SYMPOSIUM: S10 - Fundamental mechanisms in materials for next generation computing devices
CURRENT SESSION: CMOS compatible ferroelectric materials and devices
TITLE: Questions Answered and Still Outstanding Regarding the Ferroelectric Performance of Hafnia
PRESENTATION TYPE: Invited (by Invitation Only)
PRESENTER (FIRST NAME ONLY): Jon
PRESENTER (LAST NAME ONLY): Ihlefeld
PRESENTER (INSTITUTION ONLY): University of Virginia
PRESENTER (EMAIL ONLY): jfi4n@virginia.edu
PRESENTER (COUNTRY ONLY): United States
DATE/TIME SUBMITTED: October 08, 2023, 11:10 AM
ABSTRACT BODY:
Abstract Body: Ferroelectricity in hafnium oxide reenergized the thin film ferroelectric community owing to its unprecedented compatibility with CMOS and back end of line processes. Delivering on its promise of scaled FeRAM, FeFETs, and FTJs will require solving challenges associated with achieving phase purity and increasing both polarization retention and endurance. In this presentation, the current state of understanding regarding phases that exist in scaled films, the mechanisms that influence their stability, and their evolution with use will be discussed. The use of advanced characterization methods, including synchrotron-based infrared spectroscopy, scanning transmission electron microscopy, and a suite of electrical measurements will be demonstrated and discussed with the aim of identifying phases present and their evolution during use in capacitor structures. Isolating the phases present and understanding how phase transformations occur between a host of metastable and equilibrium phases is a key step toward realizing ferroelectric hafnia in mass-produced, energy saving devices.

CONTROL ID: 4000081
CURRENT SYMPOSIUM: S1: Characterization of Structure-Property Relationships in Functional Ceramics
CURRENT SESSION: Advances in scattering, imaging, and analytical techniques
TITLE: Cryogenic electron microscopy at the atomic scale: Applications and developments
PRESENTATION TYPE: Invited (by Invitation Only)
PRESENTER (FIRST NAME ONLY): Ismail
PRESENTER (LAST NAME ONLY): El Baggari
PRESENTER (INSTITUTION ONLY): Harvard University
PRESENTER (EMAIL ONLY): ie52@cornell.edu
PRESENTER (COUNTRY ONLY): United States
DATE/TIME SUBMITTED: October 08, 2023, 08:15 PM
ABSTRACT BODY:
Abstract Body: In many oxide ceramics, quantum-mechanical effects and strong electron-electron interactions give rise to superb electronic properties and a vast potential for future technologies. In these materials, electrons may self-organize into new spatial patterns that break the symmetry of the underlying crystal. These charge-ordered states display a rich interplay between distinct degrees of freedom and may induce novel functionalities through the breaking of additional crystal symmetries. I will show vivid atomic-scale visualizations of electronic order enabled by the development of cryogenic capabilities (near 100 K) for scanning transmission electron microscopy. These measurements are combined with advanced analysis tools to reveal picoscale atomic displacements governing electronic transitions, the nature and symmetry of electronic order at the local scale, and nanoscale fluctuations that underlie macroscopic properties. Despite these advances, a persistent challenge in the field of cryogenic transmission electron microscopy is the lack of stable liquid-helium-cooled sample conditions. I will discuss recent developments that provide access to ultra-low temperatures, as low as 20 Kelvin, while maintaining atomic-resolution imaging performance, paving the way for broader explorations of ultra-low temperature phenomena in oxide ceramics and beyond.

CONTROL ID: 4000112

CURRENT SYMPOSIUM: S7 - Defects and Transport in Ceramics

CURRENT SESSION: Point defect segregation to or depletion from dislocations, surfaces, grain boundaries, and interfaces

TITLE: Differentiating Surface and Bulk Defect Concentrations in Ceria-Zirconia Using Depth-Resolved XANES

PRESENTATION TYPE: Invited (by Invitation Only)

PRESENTER (FIRST NAME ONLY): Sossina

PRESENTER (LAST NAME ONLY): Haile

PRESENTER (INSTITUTION ONLY): Northwestern University

PRESENTER (EMAIL ONLY): sossina.haile@northwestern.edu

PRESENTER (COUNTRY ONLY): United States

DATE/TIME SUBMITTED: October 08, 2023, 04:18 PM

ABSTRACT BODY:

Abstract Body: Despite the immense importance of ceria-zirconia solid solutions in heterogeneous catalysis and the general consensus that catalytic activity correlates with surface defect concentrations, the surface Ce^{3+} and oxygen vacancy concentrations in these materials are unknown. Using X-ray Absorption Near Edge Spectroscopy, we quantify, under technologically relevant conditions, the Ce^{3+} concentration in the surface (2-3 nm) and bulk regions of ceria-zirconia films (~250 nm thick) grown on single-crystal yttria-stabilized zirconia. In all cases, we observe substantial Ce^{3+} enrichment at the surface. The measurements further suggest an abrupt rather than graduate decline in Ce^{3+} concentration towards the bulk value. The trends with Zr concentration are even more surprising. It is well known that the bulk Ce^{3+} concentration and hence oxygen vacancy concentration increase upon introduction of Zr. Here, we find that, although surface reduction is enhanced for all compositions, the difference between the surface and bulk regions decreases monotonically with Zr concentration. Thus, the fraction of cations in the 3+ state at the surface of the materials drops steeply as Zr is introduced. These results suggest that high Zr concentration, which is known to improve thermal stability and bulk oxygen storage capacity, may be detrimental to the local catalytic reaction rate.

CONTROL ID: 4000195
CURRENT SYMPOSIUM: S3 - Advanced Electronic Materials: Processing and Structures and Properties and Applications
CURRENT SESSION: Advanced electronic materials, including ferroelectric, piezoelectric, dielectric, electrostrictive, electrocaloric and pyroelectric materials
TITLE: Perovskite Formation Mechanisms of a Low-Lead and High Ferroelectric/Piezoelectric Property Material through High Temperature In Situ X-Ray Diffraction
PRESENTATION TYPE: Poster
PRESENTER (FIRST NAME ONLY): Grace
PRESENTER (LAST NAME ONLY): Kiel
PRESENTER (INSTITUTION ONLY): North Carolina State University
PRESENTER (EMAIL ONLY): gskiel@ncsu.edu
PRESENTER (COUNTRY ONLY): United States
DATE/TIME SUBMITTED: October 09, 2023, 08:47 PM
ABSTRACT BODY:
Abstract Body: A new ferroelectric material with the chemical composition $(1\text{-}x)Bi(Fe_{2/8}Ti_{3/8}Mg_{3/8})O_3\text{-}(x)PbTiO_3$ (BFTM-PT) is of interest due to enhanced piezoelectric properties with a high Curie temperature above 600 °C. The best properties to date have been recorded from material synthesized from a two-step cooling process, although the processing-property relationships associated with this modified thermal process are not known. The two-step process involves the starting materials being heated to 1100 °C, then cooled to 800 °C, and held for several hours before cooling to room temperature. In this work, in situ X-ray diffraction (XRD) is used to analyze phase transformations and evolution of crystalline structures during the calcination process. A comparison of the results from two different thermal processes indicate that a tetragonal structure is formed during the one-step process, and a pseudo-cubic structure is formed during the two-step process. Pseudo-cubic XRD patterns are often the result of a complex domain structure. This analysis provides evidence of the underlying mechanisms for the enhanced piezoelectric properties resulting from the two-step process. These results will inform future development of BFTM-PT and related compounds.

CONTROL ID: 4000222

CURRENT SYMPOSIUM: S2 - Advanced Microelectronics

CURRENT SESSION: Role of defects, interface and synthesis/fabrication process on device properties

TITLE: Synthesis of $Al_{1-x}Gd_xN$ wurtzite thin films using reactive HiPIMS technique

PRESENTATION TYPE: Contributed (Oral)

PRESENTER (FIRST NAME ONLY): Binod

PRESENTER (LAST NAME ONLY): Paudel

PRESENTER (INSTITUTION ONLY): Colorado School of Mines

PRESENTER (EMAIL ONLY): binod.paudel@mines.edu

PRESENTER (COUNTRY ONLY): United States

DATE/TIME SUBMITTED: October 09, 2023, 06:34 PM

ABSTRACT BODY:

Abstract Body: Alloyed AlN-based wurtzite thin films have garnered significant interest for use in ferroelectric, piezoelectric, and (opto)electronic devices. Additional functionalities may be achieved through alloying it with other elements. For example, Gd-doped $Al_{1-x}Gd_xN$ films could be excellent candidates for active components in neutron detectors because ^{157}Gd has the highest known thermal neutron capture cross-section of any element. For this and many other applications, then, it is of particular interest to explore the practical limits of alloying in such systems. Earlier work from our collaborators has predicted that compositions up to 82 cation% Gd (i.e., $Al_{0.18}Gd_{0.82}N$) should be most stable in the wurtzite structure, and our group has previously reported wurtzite films up to $Al_{0.76}Gd_{0.24}N$ via reactive RF magnetron sputtering. Here, we report our efforts to increase Gd incorporation in these single-phase wurtzite films using the higher deposition energies and self-sputtering available via High-Power Impulse Magnetron Sputtering (HiPIMS). Our HiPIMS-derived films are polycrystalline, and fiber textured, just as is typical of RF magnetron sputtered AlN-based films, but we are able to incorporate Gd contents of x>0.30. This work provides insights into deposition energetics and cation segregation processes that are applicable to the growth of other AlN-based alloy films as well.

CONTROL ID: 4000259

CURRENT SYMPOSIUM: S5 - Controlling Grain Boundary Structure, Chemistry, and their Network as a Function of Mtl Processing

CURRENT SESSION: Microstructure evolution

TITLE: Grain boundary segregation in doped $SrTiO_3$ – Impact on space charge, grain growth and conductivity

PRESENTATION TYPE: Contributed (Oral)

PRESENTER (FIRST NAME ONLY): Marc

PRESENTER (LAST NAME ONLY): Zahler

PRESENTER (INSTITUTION ONLY): Forschungszentrum Juelich

PRESENTER (EMAIL ONLY): m.zahler@fz-juelich.de

PRESENTER (COUNTRY ONLY): Germany

DATE/TIME SUBMITTED: October 09, 2023, 05:46 AM

ABSTRACT BODY:

Abstract Body: Space charge layer are known to decrease ionic conduction in perovskite-based ceramics. Besides the electrical properties space charge layer also impose a drag force during grain growth decreasing the migration rate of grain boundaries. Strontium titanate shows non-Arrhenius grain growth behavior, which results from the presence of two grain boundary types that have either fast or slow migration rates. Acceptor dopants such as Fe are known to segregate at the grain boundaries of $SrTiO_3$ and suppress the motion of fast-moving grain boundaries by solute drag. Fe-doping of $SrTiO_3$ not only impacts grain growth but also total conductivity of $SrTiO_3$.

This study focuses on doping of $SrTiO_3$ with aliovalent doping elements and their impact on the microstructure evolution, space charge layer and overall electrical properties of $SrTiO_3$. Fe-doped, Nb-doped and Nb/Fe co-doped samples were prepared and annealed at different temperatures and times. Their grain growth behavior, grain boundary chemistry and electrical properties are compared to undoped $SrTiO_3$. Space charge layers are analyzed by EIS and STEM-EDS. Changes in grain growth behavior and conductivity of doped samples are correlated to changes in the space charge layers and discussed within the framework of solute drag and defect chemistry.

CONTROL ID: 4000289
CURRENT SYMPOSIUM: S10 - Fundamental mechanisms in materials for next generation computing devices
CURRENT SESSION: Fundamental materials needs for emerging computing paradigms
TITLE: Atomic Structure Transformations of C-Doped $Ge_2Sb_2Te_5$ Thin Films Using In-Situ X-Ray Scattering and Spectroscopy
PRESENTATION TYPE: Invited (by Invitation Only)
PRESENTER (FIRST NAME ONLY): Megan
PRESENTER (LAST NAME ONLY): Butala
PRESENTER (INSTITUTION ONLY): University of Florida
PRESENTER (EMAIL ONLY): mbutala@ufl.edu
PRESENTER (COUNTRY ONLY): United States
DATE/TIME SUBMITTED: October 09, 2023, 07:58 AM
ABSTRACT BODY:
Abstract Body: Phase change materials (PCMs) are characterized by their fast transformation speeds, induced by resistive heating, from the amorphous to crystalline states and their reversibility between these states. Each state has distinct electrical, optical, and thermal properties that can be applied for non-volatile memory (NVM) data storage and, potentially, neuromorphic computation. $Ge_2Sb_2Te_5$ (GST) is a well-established PCM, which is implemented as thin films into many NVM devices due to its nanosecond scale crystallization and amorphization speeds and reliable data storage up to $\approx 10^6$ cycles. Unfortunately, GST has a low crystallization temperature of ≈135 C, which reduces phase stability and compromises long-term data storage (>10 years). Alloying GST is one way to increase the crystallization temperature. Many alloying species have been considered, with carbon providing an increased crystallization temperature.

We report the atomic structure effects of incorporating carbon into GST thin films via temperature-dependent in-situ X-ray scattering experiments. Specifically, we compare the atomic structure transformations of $(Ge_2Sb_2Te_5)_{1-x}C_x$ ($x \leq 0.12$) through X-ray diffraction (XRD), pair distribution function (PDF), and X-ray absorption spectroscopy (XAS).

CONTROL ID: 4000433
CURRENT SYMPOSIUM: S10 - Fundamental mechanisms in materials for next generation computing devices
CURRENT SESSION: CMOS compatible ferroelectric materials and devices
TITLE: Infrared Signatures for Phase Identification in Hafnium Oxide Thin Films
PRESENTATION TYPE: Contributed (Oral)
PRESENTER (FIRST NAME ONLY): Samantha
PRESENTER (LAST NAME ONLY): Jaszewski
PRESENTER (INSTITUTION ONLY): Sandia National Laboratories
PRESENTER (EMAIL ONLY): stjasze@sandia.gov
PRESENTER (COUNTRY ONLY): United States
DATE/TIME SUBMITTED: October 09, 2023, 03:46 PM
ABSTRACT BODY:
Abstract Body: Phase identification in HfO_2-based thin films is necessary for understanding the mechanisms stabilizing the ferroelectric phase in these materials that are promising for next-generation non-volatile memory and computing technology. While grazing-incidence X-ray diffraction is often used for this purpose, it has difficulty unambiguously differentiating between the ferroelectric phase and other metastable phases that may exist due to similarities in the d-spacings. Infrared signatures provide an alternative route. In this work, the IR optical responses of three hafnium oxide films, where annealing conditions were used to create films consisting of different phases, are reported. Vibrational signatures acquired from synchrotron nano-Fourier transform infrared spectroscopy (nano-FTIR) are shown to be capable of differentiating between the phases in a non-destructive, rapid, and nanoscale manner. By demonstrating that IR-spectroscopy can unambiguously distinguish phases in this material, this work establishes a tool needed to isolate the factors dictating ferroelectric phase stability in HfO_2-based technologies.

CONTROL ID: 4000439
CURRENT SYMPOSIUM: S14 - Emerging Semiconductor Materials and Interfaces
CURRENT SESSION: WBG and UWBG materials - synthesis, characterization, theory
TITLE: Compositional Defects and Ferroelectric Properties of $Al_{1-x}Sc_xN$: A First Principles Study
PRESENTATION TYPE: Contributed (Oral)
PRESENTER (FIRST NAME ONLY): Victoria
PRESENTER (LAST NAME ONLY): Bradford
PRESENTER (INSTITUTION ONLY): University of Connecticut
PRESENTER (EMAIL ONLY): victoria.bradford@uconn.edu
PRESENTER (COUNTRY ONLY): United States
DATE/TIME SUBMITTED: October 09, 2023, 02:04 PM
ABSTRACT BODY:
Abstract Body: $Al_{1-x}Sc_xN$ shows notable ferroelectric characteristics with Sc concentration reaching beyond ~22% and has a potential usage in various applications like high electron mobility transistors (HEMT), micro electromechanical systems (MEMS) and optical cells. Theoretically, it is known that adding Sc into wurtzite AlN enables ferroelectric switching by reducing the energy barrier between the ferroelectric and paraelectric phases, with a very large remnant polarization and coercive field below the dielectric breakdown strength. Although it's remarkable features, increasing Sc concentration also leads to changes in electronic structure and introduces leakage current that limits the applications. These features are also related to experimental conditions, like material itself intrinsically has defects like non-metals and vacancy. These defects influence quality of the sample and the expected ferroelectric properties especially spontaneous polarization, ferroelectric switching energy and coervice field. The aim of this study is to obserce how these defects influence ferroelectric properties in $Al_{0.75}Sc_{0.25}N$ using density functional theory (DFT) calculations. We compared the ferroelectric and paraelectric phase stability under the influence of these defects and identified how their co-existence reflects on the ferroelectric properties.

CONTROL ID: 4000494
CURRENT SYMPOSIUM: S16 - New Frontiers and Science in Additive Manufacturing of Ceramic Materials
CURRENT SESSION: New approaches for AM processes of ceramics and ceramic-matrix composites
TITLE: Hydrothermal-assisted Jet Fusion: A Selective Cold Sintering Approach
PRESENTATION TYPE: Invited (by Invitation Only)
PRESENTER (FIRST NAME ONLY): Xuan
PRESENTER (LAST NAME ONLY): Song
PRESENTER (INSTITUTION ONLY): University of Iowa
PRESENTER (EMAIL ONLY): xuan-song@uiowa.edu
PRESENTER (COUNTRY ONLY): United States
DATE/TIME SUBMITTED: October 09, 2023, 04:35 PM
ABSTRACT BODY:
Abstract Body: Ceramic additive manufacturing (AM) typically uses a high fraction of organic binders to form pre-sintered green parts that require a post de-binding process to remove. The de-binding process inevitably results in severe gas expansion and residual chars, leading to structural defects, accumulated stress, and compromised material properties in the final parts. Here we report a binder-free additive manufacturing process that utilizes a hydrothermal method to create geometrically and compositionally complex ceramics under mild temperatures. The hydrothermal method utilizes a selectively deposited volatile dissolving ink, high pressure, and milt heat to strategically fuse a ceramic powder bed into complex geometries. Compared to traditional AM methods for ceramics, the HJF process eliminates the need for organic binders in green part fabrication and offers the potential to directly co-print ceramics with other dissimilar materials, such as polymers and metals, enabling the development of novel multi-functional ceramic composites.

CONTROL ID: 4000498

CURRENT SYMPOSIUM: S16 - New Frontiers and Science in Additive Manufacturing of Ceramic Materials

CURRENT SESSION: Mechanical, thermal, electrical, magnetic, and optical properties

TITLE: Additive manufacturing of bicontinuous piezocomposites based on triply periodic cellular micro-skeleton for tailorable hydrostatic performance

PRESENTATION TYPE: Contributed (Oral)

PRESENTER (FIRST NAME ONLY): Xuan

PRESENTER (LAST NAME ONLY): Song

PRESENTER (INSTITUTION ONLY): University of Iowa

PRESENTER (EMAIL ONLY): xuan-song@uiowa.edu

PRESENTER (COUNTRY ONLY): United States

DATE/TIME SUBMITTED: October 09, 2023, 04:39 PM

ABSTRACT BODY:

Abstract Body: Bicontinuous piezoelectric composites, consisting of a continuous piezoelectric ceramic skeleton and a flexible polymer phase interconnected in three-dimension (3D), possess an excellent property combination of mechanical flexibility and piezoelectricity. The piezoelectric performances of these materials are dependent on the meso-scale geometry of the interface between the active piezoelectric ceramic phase and the passive polymer phase. In this research, we investigate the design and fabrication of bicontinuous piezocomposites with tailorable hydrostatic performance utilizing a suspension enveloped projection stereolithography process (SEPS). Triply periodic cellular structures are used to design the active piezoelectric ceramic phase, due to the large elastic constants with respect to their rod-connected counterparts. Post processes including debinding, sintering, poling and polymer infiltration are presented. Mechanical and piezoelectric properties of the achieved bicontinuous piezocomposites are analyzed through experimental characterization and finite element modeling.

CONTROL ID: 4000510

CURRENT SYMPOSIUM: S13 - Frontiers in Ferroic oxides: Synthesis, Properties, and Applications

CURRENT SESSION: Advanced synthesis and characterization techniques for ferroic oxides

TITLE: Direct Imaging of Polar Topological Textures and Their Phase Transitions using Electron Microscopy

PRESENTATION TYPE: Invited (by Invitation Only)

PRESENTER (FIRST NAME ONLY): Yu-Tsun

PRESENTER (LAST NAME ONLY): Shao

PRESENTER (INSTITUTION ONLY): University of Southern California

PRESENTER (EMAIL ONLY): yutsunsh@usc.edu

PRESENTER (COUNTRY ONLY): United States

DATE/TIME SUBMITTED: October 09, 2023, 06:50 PM

ABSTRACT BODY:

Abstract Body: Emerging topological structures like polar vortices, skyrmions, and merons in ferroelectric heterostructures have garnered significant attention due to their intricate underlying physics and prospective applications in future nanodevices. While concepts such as topological protection and phase transitions influenced by temperature, strain, electric field, and superlattice periodicity have been explored in polar topological defects like vortices, investigations into polar topological solitons, such as skyrmions and merons, remain in its infancy. Understanding these transitions requires a characterization method which can track the detailed 3D structure of individual dipole textures. By integrating four-dimensional phase space information recorded using direct electron detectors (4D-STEM) with dynamical diffraction simulations and ptychographic phase retrieval algorithms, we have been able to robustly image the local polarization and chirality within each polar texture.

In this talk, I will show how 4D-STEM-based approach allows us to image dipolar waves, polar skyrmions and merons in ferroelectric thin films that are unable to measure by conventional methods. Furthermore, we study the topological phase transitions among these polar textures as perturbed by external stimuli such as strain and electric-field gradient generated by the focused electron probe.

CONTROL ID: 4000522
CURRENT SYMPOSIUM: S1: Characterization of Structure-Property Relationships in Functional Ceramics
CURRENT SESSION: Addressing open questions in functional ceramics
TITLE: Constraining the Size of Chemical Nanoregions in Barium Zriconate-Titanate Relaxors
PRESENTATION TYPE: Contributed (Oral)
PRESENTER (FIRST NAME ONLY): Charles
PRESENTER (LAST NAME ONLY): Evans
PRESENTER (INSTITUTION ONLY): Carnegie Mellon University
PRESENTER (EMAIL ONLY): charlese@andrew.cmu.edu
PRESENTER (COUNTRY ONLY): United States
DATE/TIME SUBMITTED: October 09, 2023, 09:18 PM
ABSTRACT BODY:
Abstract Body: Barium zirconate-titanate (BZT) is a lead-free relaxor material. Relaxors are ferroelectric-like materials which exhibit only short-range polar order, rather than the long-range order observed in traditional ferroelectric materials. The presence of Zr4+ ions in BZT disrupts the ferroelectric displacement of nearby Ti4+ ions, inhibiting the formation of long-range polar ordering usually observed in barium titanate.

Several studies of relaxor BZT using techniques such as extended x-ray absorption fine structure and Raman spectroscopy have found evidence suggesting the presence of "chemical nanoregions" enriched in Zr compared to the bulk concentration. This work has attempted to directly observe Zr-enriched nanoregions in BZT via HAADF scanning transmission electron microscopy (STEM), quantified using spatial statistics methods. No Zr-enriched nanoregions have been detected. With the aid of simulated STEM images this work now attempts to constrain the possible scale of chemical nanoregions in BZT in light of their experimental non-detection via STEM. At time of submission further work using atom-probe tomography is being planned, with the aim to add three dimensional information to augment existing two dimensional STEM data.

CONTROL ID: 4000531
CURRENT SYMPOSIUM: S2 - Advanced Microelectronics
CURRENT SESSION: Memristive switching and performance
TITLE: Origins of Information Retention in Memristive Materials
PRESENTATION TYPE: Contributed (Oral)
PRESENTER (FIRST NAME ONLY): Yiyang
PRESENTER (LAST NAME ONLY): Li
PRESENTER (INSTITUTION ONLY): University of Michigan
PRESENTER (EMAIL ONLY): yiyangli@umich.edu
PRESENTER (COUNTRY ONLY): United States
DATE/TIME SUBMITTED: October 09, 2023, 06:00 PM
ABSTRACT BODY:
Abstract Body: Redox-based resistive memory devices, also known as memristors or ReRAM, are highly promising elements for embedded nonvolatile memory, in-memory computing, and neuromorphic computing. Such devices switch resistance states through the electrochemical migration of oxygen vacancies in transition metal oxides. We present our recent research on the materials thermodynamics of ionic motion in these oxygen-based memristors. Using a combination of device measurements, materials characterization, and multiscale physical modeling, we find that oxygen vacancies do not obey Fick's First Law of diffusion as conventionally believed, but instead undergo composition phase separation, which enables diffusion against the concentration gradient. This phase separation is critical to the ability of memristors to retain information for long periods of time. Finally, we utilize this understanding of phase separation in transition metal oxides to engineer retention time in a new class of three-terminal electrochemical memory.

CONTROL ID: 4000532
CURRENT SYMPOSIUM: S1: Characterization of Structure-Property Relationships in Functional Ceramics
CURRENT SESSION: Advances in scattering, imaging, and analytical techniques
TITLE: Monochromated and Aberration-corrected STEM for Quantum Materials
PRESENTATION TYPE: Invited (by Invitation Only)
PRESENTER (FIRST NAME ONLY): Andrew
PRESENTER (LAST NAME ONLY): Lupini
PRESENTER (INSTITUTION ONLY): Oak Ridge National Lab
PRESENTER (EMAIL ONLY): arl1000@ornl.gov
PRESENTER (COUNTRY ONLY): United States
DATE/TIME SUBMITTED: October 09, 2023, 08:50 PM
ABSTRACT BODY:
Abstract Body: Quantum materials" display effects that cannot be well explained by classical physics descriptions. Although other tools provide macroscopic or bulk methods to probe these effects, scanning transmission electron microscopy (STEM) is one of the foremost tools for local analysis of quantum materials at high-spatial resolution in real space.

Over the last two decades, the intrinsic electron-optical aberrations that limit the spatial resolution have been improved by aberration-correction, and now monochromation is able to improve upon the intrinsic energy spread of the electron source. Monochromated, aberration-corrected STEM now provides a combination of atomic-scale spatial resolution and few-meV energy resolution. However, challenges remain to view quantum materials at the relevant scale. In particular, many quantum effects only happen at low-temperatures, placing new demands upon stage and microscope design and stabilities. One of the strengths of the STEM is that as well as providing a tool to study these materials, a STEM provides potential routes to manipulate and perhaps create novel quantum structures. STEM is able to image and also to change structures at the atomic-scale. However, some of these processes are still poorly understood and lack fine control. The ability to locally heat and excite samples, or to add dopants with single atom control will provide exciting future opportunities.

CONTROL ID: 4000542
CURRENT SYMPOSIUM: S7 - Defects and Transport in Ceramics
CURRENT SESSION: Defect-mediated properties
TITLE: Polaron formation: A bottleneck in photoexcited electrons' transport in metal-oxides
PRESENTATION TYPE: Contributed (Oral)
PRESENTER (FIRST NAME ONLY): Muhammad
PRESENTER (LAST NAME ONLY): Huda
PRESENTER (INSTITUTION ONLY): University of Texas at Arlington
PRESENTER (EMAIL ONLY): huda@uta.edu
PRESENTER (COUNTRY ONLY): United States
DATE/TIME SUBMITTED: October 09, 2023, 07:25 PM
ABSTRACT BODY:
Abstract Body: Functional meta-oxides, such as those intended for solar-energy conversion processes, are required to absorb sunlight, excite electrons, and transport the photo-excited charge carriers efficiently. The overall energy conversion efficiencies of these photo-active materials are not as high as expected, even with near-suitable bandgap energies required for efficient solar energy absorptions. One of the reasons is attributed to the poor charge transport properties in metal-oxides having strongly correlated electrons. Typically, strongly correlated systems exhibit polaron conduction via electron-phonon interactions. Recently, there have been significant efforts to characterize polaronic states and their effects on transport properties in photoactive materials. First-principles density functional theory (DFT) studies can reveal the electronic origin of the polaron formation and, therefore, can suggest how to manipulate the overall transport properties in such photoactive materials. In this presentation, representative metal-oxides photo-catalysts will be considered to demonstrate the small polaron formation mechanism. The polaronic states in metal-oxides will be depicted from detailed electronic structure calculations by DFT based methods. We will show how these polaronic states affect the optical band gaps. In addition, a strategy will be shown that can inhibit polaron formations in meral-oxides.

CONTROL ID: 4000544

CURRENT SYMPOSIUM: S3 - Advanced Electronic Materials: Processing and Structures and Properties and Applications

CURRENT SESSION: Advanced electronic materials, including ferroelectric, piezoelectric, dielectric, electrostrictive, electrocaloric and pyroelectric materials

TITLE: Charting The Application Space for Textured PMN-PZ-PT Piezoceramics

PRESENTATION TYPE: Poster

PRESENTER (FIRST NAME ONLY): Ahmet

PRESENTER (LAST NAME ONLY): Gurdal

PRESENTER (INSTITUTION ONLY): QorTek, Inc.

PRESENTER (EMAIL ONLY): egurdal@qortek.com

PRESENTER (COUNTRY ONLY): United States

DATE/TIME SUBMITTED: October 09, 2023, 06:59 PM

ABSTRACT BODY:

Abstract Body: The degree of crystallographic orientation is a key factor to achieve high performance in piezoelectric ceramics. This effect has been well documented over a few decades via studies in single crystal and textured piezoceramics. Recently, the latter gained quite a bit of interest from the researchers from academia and the industry due to its manufacturability and cost advantages over the former. However, there remain obstacles to a more widespread market acceptance of textured piezoceramics as a high-performance alternative to legacy PZT materials and devices. This is because of the limited application relevant properties of the textured piezoceramics and most transducer designers being hesitant to go into uncharted territories for their design updates or next generation devices.

Here we explore the temperature, electric field, and stress dependence of the commercially available textured PMN-PZ-PT piezoceramic (TX101; d_{33}: 710 pC/N, k_p: 0.75, tan d: 0.003, and Q_m: 350) to provide more insight into these uncharted territories. The results show that TX101 is functional between -196C to 150C, can be driven up to 480 V_{rms} /mm electric field, and has comparable mechanical properties to PZT.

CONTROL ID: 4000545

CURRENT SYMPOSIUM: S4 - Complex Oxide Thin Films and Heterostructures: From Synthesis to Strain/Interface-engineered Emergent Properties

CURRENT SESSION: Synthesis and properties of high entropy complex oxides

TITLE: Unraveling the Strain Effect on Oxygen Evolution Reaction Activities of A-site High Entropy Nickelates

PRESENTATION TYPE: Contributed (Oral)

PRESENTER (FIRST NAME ONLY): Min-Ju

PRESENTER (LAST NAME ONLY): Choi

PRESENTER (INSTITUTION ONLY): Pacific Northwest National Laboratory

PRESENTER (EMAIL ONLY): minju.choi@pnnl.gov

PRESENTER (COUNTRY ONLY): United States

DATE/TIME SUBMITTED: October 09, 2023, 10:39 PM

ABSTRACT BODY:

Abstract Body: High-entropy oxides (HEOs) have attracted great attention as an emerging material system for the development of highly active electrocatalysts due to coexistence of multiple potential active sites and the inherent tunability. However, the inhomogeneous surface composition, ill-defined crystallographic surface phases, and exposed facets of bulk HEOs make it difficult to extract intrinsic activity and stability trends. In our study, we have synthesized a set of A-site high entropy perovskite nickelate $La_{0.2}Nd_{0.2}Pr_{0.2}Sm_{0.2}Eu_{0.2}NiO_{3-\delta}$ (HEN) epitaxial thin films on different single crystal substrates and investigated the influence of strain on their crystal structures, electronic behaviors, and oxygen evolution reaction (OER) activities. We demonstrate that HEN films exhibit slightly higher OER activities compared to pure $NdNiO_3$ films grown on the same substrate, with the highest OER activity observed for the HEO film grown on the LSAT substrate. Detailed analysis using X-ray absorption spectroscopy reveals a correlation between the OER activity of the HEN films and the intensity of O K pre-edge peak. Additionally, we find that both HEO and pure $NdNiO_3$ films exhibit similar trends in OER stability. This research sheds light on the potential of HEOs as catalysts and their response to strain, offering valuable insights for the design of high-performance electrocatalytic materials.

CONTROL ID: 4000548
CURRENT SYMPOSIUM: S16 - New Frontiers and Science in Additive Manufacturing of Ceramic Materials
CURRENT SESSION: New approaches for AM processes of ceramics and ceramic-matrix composites
TITLE: High-throughput, Ultra-fast Laser Sintering of Ceramics and Machine-learning-Based Prediction on Processing-Microstructure-Property Relationships
PRESENTATION TYPE: Invited (by Invitation Only)
PRESENTER (FIRST NAME ONLY): Fei
PRESENTER (LAST NAME ONLY): Peng
PRESENTER (INSTITUTION ONLY): Clemson University
PRESENTER (EMAIL ONLY): fpeng@clemson.edu
PRESENTER (COUNTRY ONLY): United States
DATE/TIME SUBMITTED: October 09, 2023, 07:02 PM
ABSTRACT BODY:
Abstract Body: We report high-throughput, ultra-fast laser sintering of alumina sample array and characterization of sample units' microstructure and hardness, as a fast exploration of laser processing parameters, microstructure, and property. These experimental data were used to train machine-learning (ML) models. Accurate ML predictions were demonstrated for the processing-microstructure-property relationship, specifically in (1) prediction of the microstructure of alumina under arbitrary laser power and (2) prediction of the expected microstructure from the desired hardness. An independent neural network was developed and showed that ML-predicted microstructure had less than 10% error from real ones, in terms of projected hardness. To monitor the microstructure during laser sintering, we demonstrated an ML model that can instantaneously predict ceramic's microstructure at the laser spot, based on the laser spot brightness. The ML model can generate more than 10 predictions per second, and the error in average grain size was less than 5% from the experimental observations.

CONTROL ID: 4000558
CURRENT SYMPOSIUM: S16 - New Frontiers and Science in Additive Manufacturing of Ceramic Materials
CURRENT SESSION: New approaches for AM processes of ceramics and ceramic-matrix composites
TITLE: Fabrication and characterization of CO_2 laser-sintered silica-titania glass
PRESENTATION TYPE: Contributed (Oral)
PRESENTER (FIRST NAME ONLY): Siddhartha
PRESENTER (LAST NAME ONLY): Sarkar
PRESENTER (INSTITUTION ONLY): Clemson University
PRESENTER (EMAIL ONLY): sarkar2@g.clemson.edu
PRESENTER (COUNTRY ONLY): United States
DATE/TIME SUBMITTED: October 09, 2023, 07:35 PM
ABSTRACT BODY:
Abstract Body: Silica-titania glass is a ubiquitous material in optics and photonics, owing to its extremely low coefficient of thermal expansion, spectral range in the visible and near IR regions and a high refractive index. Techniques such as FDM, SLA, SLS, sol-gel based dip-coating and DIW have been used to obtain transparent silica-titania films or bulk structures. However, these techniques are time-consuming and require elaborate post-processing steps to obtain the desired product. In this study, a novel CO_2 laser-based sintering approach is demonstrated to fabricate a thin layer of silica-titania glass on fused silica substrates. The addition of different weight percentages of TiO_2 are explored to find out the extent to which TiO_2 can be dissolved in the silica glass network. The laser-sintered coatings were crack-free and transparent up to a certain weight percentage of TiO_2, after which devitrification of the silica network and thermochromism in TiO_2 took over. Marangoni effect plays a dominant role in laser sintering of glass here and gives rise to the characteristic pattern and pore removal mechanism. The laser-sintered glass coatings are characterized using Raman, XRD, UV-VIS and surface profilometer to gauge the surface profile and roughness. The results agree well with established literature and is potentially a novel method to fabricate intricate glass patterns in the photonics and optics industries.

CONTROL ID: 4000560
CURRENT SYMPOSIUM: S13 - Frontiers in Ferroic oxides: Synthesis, Properties, and Applications
CURRENT SESSION: Ferroic materials for novel devices
TITLE: Non-volatile electric-field control of inversion symmetry in a multiferroic oxide
PRESENTATION TYPE: Invited (by Invitation Only)
PRESENTER (FIRST NAME ONLY): Lucas
PRESENTER (LAST NAME ONLY): Caretta
PRESENTER (INSTITUTION ONLY): Brown University
PRESENTER (EMAIL ONLY): lucas_caretta@brown.edu
PRESENTER (COUNTRY ONLY): United States
DATE/TIME SUBMITTED: October 09, 2023, 07:42 PM
ABSTRACT BODY:
Abstract Body: Competition between ground states at phase boundaries can lead to significant changes in properties under stimuli, particularly when these ground states have different crystal symmetries. A key challenge is to stabilize and control the coexistence of symmetry-distinct phases. Using $BiFeO_3$ layers confined between layers of dielectric $TbScO_3$ as a model system, we stabilize the mixed-phase coexistence of centrosymmetric and non-centrosymmetric $BiFeO_3$ phases at room temperature with antipolar, insulating and polar semiconducting behaviour, respectively. Application of orthogonal in-plane electric (polar) fields results in reversible non-volatile interconversion between the two phases, hence removing and introducing centrosymmetry. Counterintuitively, we find that an electric field 'erases' polarization, resulting from the anisotropy in octahedral tilts introduced by the interweaving $TbScO_3$ layers. Consequently, this interconversion between centrosymmetric and non-centrosymmetric phases generates changes in the non-linear optical response of over three orders of magnitude, resistivity of over five orders of magnitude and control of microscopic polar order. Our work establishes a platform for cross-functional devices that take advantage of changes in optical, electrical and ferroic responses, and demonstrates octahedral tilts as an important order parameter in materials interface design.

CONTROL ID: 4000562
CURRENT SYMPOSIUM: S4 - Complex Oxide Thin Films and Heterostructures: From Synthesis to Strain/Interface-engineered Emergent Properties
CURRENT SESSION: Theory, modeling and first principles calculations of complex oxides and their heterostructures
TITLE: Oxygen vacancy ordering and oxygen transport in topotactic oxides
PRESENTATION TYPE: Invited (by Invitation Only)
PRESENTER (FIRST NAME ONLY): Peter
PRESENTER (LAST NAME ONLY): Sushko
PRESENTER (INSTITUTION ONLY): Pacific Northwest National Lab
PRESENTER (EMAIL ONLY): peter.sushko@pnnl.gov
PRESENTER (COUNTRY ONLY): United States
DATE/TIME SUBMITTED: October 10, 2023, 02:56 AM
ABSTRACT BODY:
Abstract Body: Properties of metal oxides and metal oxide/metal heterojunctions are strongly dependent on the concentration and spatial distribution of oxygen species and, conversely, of oxygen vacancies. Controlling these properties requires an understanding of the atomic scale factors that define the stability of these oxygen species, their transport properties, and their effect on the electronic structure of the surrounding lattice. Here we use ab initio simulations to obtain an atomic-level insight into the coupling between chemical composition, oxygen vacancy content, and the lattice strain in topotactic metal oxides $SrFeO_{3-x}$, $SrCrO_{3-x}$, and their solid solutions and heterostructures $Sr(Fe,Cr)O_{3-x}$. We also discuss the effect of this coupling of the atomic transport, including the dependence of oxygen diffusion mechanisms on the spatial ordering of the oxygen vacancies and opportunities for exploiting this ordering to create new phases of sub-stoichiometric oxides.

Z. Yang, L. Wang, J. A. Dhas, M. H. Engelhard, M. E. Bowden, W. Liu, Z. Zhu, C. Wang, S. A. Chambers, P. V. Sushko, Y. Du, Guided anisotropic oxygen transport in vacancy ordered oxides, Nature Communications 14, 6068 (2023). DOI: 10.1038/s41467-023-40746-4

CONTROL ID: 4000567
CURRENT SYMPOSIUM: S5 - Controlling Grain Boundary Structure, Chemistry, and their Network as a Function of Mtl Processing
CURRENT SESSION: Mechanical and electric fields (SPS, FAST, HIP, etc.)
TITLE: Bridging Mesoscale and Engineering Scale in Pressure-assisted Sintering Modeling
PRESENTATION TYPE: Contributed (Oral)
PRESENTER (FIRST NAME ONLY): Danny
PRESENTER (LAST NAME ONLY): Hermawan
PRESENTER (INSTITUTION ONLY): Purdue University
PRESENTER (EMAIL ONLY): dhermawa@purdue.edu
PRESENTER (COUNTRY ONLY): United States
DATE/TIME SUBMITTED: October 09, 2023, 10:16 PM
ABSTRACT BODY:
Abstract Body: A fundamental understanding on the microstructural evolution during sintering is key in tailoring the properties of ceramics for a wide range of applications, including structural, energy storage, and electronic devices. Significant contributions to mass transport and interfacial motion are local elastic stresses and plastic flow, driving neck formation and grain boundary motion at the particle-particle contacts. These granular geometries are often linked to the densification process, resulting in complex stress distributions, and challenging the classic relationship of porosity and local stress at the engineering scale. In this presentation, a combination of data analytics and phase-field modeling was develop to: first, model the microstructural evolution in combination with large deformation mechanics; second, investigate the relationship between local mechanical response and kinetic laws of motion using data analytics; and third, establish a relationship between the mesoscale and the engineering scale to model the coarse grained densification process.

CONTROL ID: 4000574
CURRENT SYMPOSIUM: S3 - Advanced Electronic Materials: Processing and Structures and Properties and Applications
CURRENT SESSION: Materials design, new materials and structures, and their emerging applications
TITLE: Deactivating the lower level in Ho/Pr:Y_2O_3 ceramic and activating the upper level in Er/Yb:Y_2O_3 ceramic for mid-IR lasers
PRESENTATION TYPE: Poster
PRESENTER (FIRST NAME ONLY): Yuhao
PRESENTER (LAST NAME ONLY): Zhan
PRESENTER (INSTITUTION ONLY): East China Normal University
PRESENTER (EMAIL ONLY): 51215904063@stu.ecnu.edu.cn
PRESENTER (COUNTRY ONLY): China
DATE/TIME SUBMITTED: October 09, 2023, 10:37 PM
ABSTRACT BODY:
Abstract Body: This article investigates the comparison on the fluorescence spectra, upconversion spectra and the fluorescence lifetimes of Ho^{3+} transitions $^5I_6 \rightarrow {}^5I_8$, $^5I_7 \rightarrow {}^5I_8$ and $^5I_6 \rightarrow {}^5I_7$ and Er^{3+} transitions $^4I_{13/2} \rightarrow {}^4I_{15/2}$ and $^4I_{11/2} \rightarrow {}^4I_{13/2}$ for Ho:Y_2O_3 with and without co-doping of Pr^{3+} ions and Er:. Y_2O_3 with and without co-doping of Yb^{3+} ions, respectively. Accordingly, we analyze the population variation on both upper and lower laser levels under the influence of fluorescence emission, upcoversion luminescence and multiple energy transfers. We obtain the enhancement of mid-IR 2.845 μm and 2.722 μm emissions through the co-doping of Pr^{3+} into Ho:Y_2O_3 and Er^{3+} into Er:. Y_2O_3, respectively.

CONTROL ID: 4000589
CURRENT SYMPOSIUM: S15 - Electronics Materials for 5G/6G Applications
CURRENT SESSION: Materials growth and design
TITLE: Leakage reduction and polarization saturation in Al0.7Sc0.3N films integrated on GaN and InGaN templates
PRESENTATION TYPE: Contributed (Oral)

PRESENTER (FIRST NAME ONLY): Keisuke
PRESENTER (LAST NAME ONLY): Yazawa
PRESENTER (INSTITUTION ONLY): Colorado School of Mines
PRESENTER (EMAIL ONLY): Keisuke.Yazawa@nrel.gov
PRESENTER (COUNTRY ONLY): United States

DATE/TIME SUBMITTED: October 09, 2023, 09:52 PM

ABSTRACT BODY:

Abstract Body: Wurtzite ferroelectrics are a promising candidate for microelectronics applications due to their process and crystallographic compatibility to tetrahedrally bonded semiconductors such as Si and GaN. Of particular interest is the (Al,Sc)N stack on GaN because of its close lattice matching and immediate application for a ferroelectric HEMT. Although previous studies have successfully observed ferroelectric switching in (Al,Sc)N films on GaN, the challenge lies in mitigating the leakage current associates with the (Al,Sc)N films, which can result in undesired charge injection through the (Al,Sc)N gate within the HEMT device.

We demonstrate a reduction of the leakage current contribution during the ferroelectric switching process in epitaxial $Al_{0.7}Sc_{0.3}N$ films on GaN and $In_{0.2}Ga_{0.8}N$ templated substrates. The epitaxial relation is confirmed by X-ray diffraction pole figures and transmission electron microscopy. The polarization – electric field hysteresis loops are well-saturated, indicating that the leakage current contribution to the total accumulated charges during the external voltage sequence is insignificant compared to the switching current contribution. The low-loss $Al_{0.7}Sc_{0.3}N$ film integrated on a GaN HEMT device exhibits unambiguous ferroelectric switching.

CONTROL ID: 4000600

CURRENT SYMPOSIUM: S17 - Electronic and Ionic Materials in Battery Energy Storage

CURRENT SESSION: Next Gen Electrolytes

TITLE: Improving the ionic conductivity of NaTiGaO electrolytes by pairing fast sintering, grain growth analysis, high throughput measurements, and data-driven computer models

PRESENTATION TYPE: Contributed (Oral)

PRESENTER (FIRST NAME ONLY): Michael

PRESENTER (LAST NAME ONLY): Thuis

PRESENTER (INSTITUTION ONLY): Northwestern University

PRESENTER (EMAIL ONLY): michaelthuis2026@u.northwestern.edu

PRESENTER (COUNTRY ONLY): United States

DATE/TIME SUBMITTED: October 09, 2023, 10:43 PM

ABSTRACT BODY:

Abstract Body: Rapidly increasing demand for electric vehicles is straining the supply of battery materials. This work seeks to develop novel materials for improved batteries using solid-state electrolytes and alternative battery chemistries by combining high throughput electrochemical measurements and data analysis with computational models. These methods are paired with an ultra-fast high-temperature synthesis (UHS) method allowing access to improved compositions with simplified manufacturing steps. Optimized compositions of NaTiGaO compounds were identified as good ion conductors by the computational model. Targeting the optimized stoichiometry improved bulk ionic conductivity by ~25% from an initial conductivity of 0.137 mS/cm. The UHS method retained more sodium than traditional ceramic solid-state synthesis, allowing a further ~50% increase in bulk ionic conductivity. Scanning transmission electron microscopy (STEM) analysis is used to understand these materials' unique grain growth and microstructure under fast sintering conditions. STEM observed complex phase boundaries and morphologies contributing to poor grain boundary ionic conductivity performance.

CONTROL ID: 4000604
CURRENT SYMPOSIUM: S5 - Controlling Grain Boundary Structure, Chemistry, and their Network as a Function of Mtl Processing
CURRENT SESSION: Grain growth and mobility
TITLE: Anisotropic Grain Growth in Polycrystalline Films
PRESENTATION TYPE: Contributed (Oral)
PRESENTER (FIRST NAME ONLY): Danny
PRESENTER (LAST NAME ONLY): Hermawan
PRESENTER (INSTITUTION ONLY): Purdue University
PRESENTER (EMAIL ONLY): dhermawa@purdue.edu
PRESENTER (COUNTRY ONLY): United States
DATE/TIME SUBMITTED: October 09, 2023, 10:24 PM
ABSTRACT BODY:
Abstract Body: Understanding the fundamental physics of grain boundary motion to tailor the microstructure of polycrystalline thin films for novel semiconducting and energy applications is key to improve its production and functionalities. N. Thangaraj and U. Dahmen [Mat. Res. Soc. Symp. Proc. 238:171-176, 1992.] detailed the grain morphology and interfacial structure of a <100> orientation, heteroepitaxial Aluminum thin film grains network on a Si (111) substrate. A closer look on the three-crystallograpic orientation cellular network develops during the PVD growth and annealing process shows grain structures and dihedral angles that differ from the traditional 120 degrees, and calls attention to the relationship between local crystallographic anisotropy of interfacial energy and the five-dimensional space of geometrical misorientation parameters that describe a grain boundary. In this presentation, a combined data analytics, sharp and thermodynamically-consistent diffuse interface formulations was utilized to develop detailed insight into the relationship between interfacial thermodynamics and kinetic laws of motions for grain structures with anisotropic interfacial energy.

CONTROL ID: 4000620
CURRENT SYMPOSIUM: S1: Characterization of Structure-Property Relationships in Functional Ceramics
CURRENT SESSION: Advances in scattering, imaging, and analytical techniques
TITLE: In-situ Air-Free 4D-STEM Biasing of Model Batteries
PRESENTATION TYPE: Contributed (Oral)
PRESENTER (FIRST NAME ONLY): Hadas
PRESENTER (LAST NAME ONLY): Sternlicht
PRESENTER (INSTITUTION ONLY): National Center for Electron Microscopy, Lawrence Berkeley National Laboratory
PRESENTER (EMAIL ONLY): hsternlicht@lbl.gov
PRESENTER (COUNTRY ONLY): United States
DATE/TIME SUBMITTED: October 09, 2023, 11:47 PM
ABSTRACT BODY:
Abstract Body: While sulfur has been reported as a promising cathode, practical use of lithium-sulfur batteries is impeded by challenges such as the low electric conductivity of sulfur. This work determines microstructural changes in lithium-sulfur batteries upon biasing, which is crucial for identifying the mechanisms that alter lithium-ion transport and affect battery performance. To do so, Lithium-Sulfur micrometer length-scale solid state batteries are fabricated, biased and studied entirely in-situ using 4D-STEM. This approach allows focus on a "single" controllable interface. A full air-free setup was developed and is being used to transfer samples to and from the glove box during sample preparation and characterization. Carbon is introduced to and coated on the sulfur cathode, in order to exploit the improved properties associated with the sulfur(carbon) cathode compared to sulfur. These include improved conductivity, and prevention of sulfur loss via vapor pressure.

CONTROL ID: 4000621

CURRENT SYMPOSIUM: S3 - Advanced Electronic Materials: Processing and Structures and Properties and Applications

CURRENT SESSION: Advanced electronic materials, including ferroelectric, piezoelectric, dielectric, electrostrictive, electrocaloric and pyroelectric materials

TITLE: Wurtzite Ferroelectric Heterostructures in Oxide and Nitride Systems

PRESENTATION TYPE: Contributed (Oral)

PRESENTER (FIRST NAME ONLY): Chloe

PRESENTER (LAST NAME ONLY): Skidmore

PRESENTER (INSTITUTION ONLY): Pennsylvania State University

PRESENTER (EMAIL ONLY): chs5204@psu.edu

PRESENTER (COUNTRY ONLY): United States

DATE/TIME SUBMITTED: October 09, 2023, 11:38 PM

ABSTRACT BODY:

Abstract Body: This presentation will demonstrate ferroelectric heterostructures in the $Zn_{1-x}Mg_xO$, $Al_{1-x}Sc_xN$, and $Al_{1-x}B_xN$ solid solution systems. In each case thin film stacks that include pure ZnO and AlN films are stacked with their compositionally modified solid solution companions. In so doing, the entire stack is switchable, i.e., polarization reversal occurs in both the modified and the pure layers. Proof of switching is demonstrated by hysteresis measurements of parallel plate capacitors, by piezo-force microscopy before and after switching, by chemical etching anisotropy of polarization-up and polarization-down patterns, and by direct polarization measurement from high-resolution cross-sectional STEM images. The ability to switch conventionally non-ferroelectric ZnO and AlN layers is demonstrated in symmetric (ZMO-ZnO-ZMO and AlBN-AlN-AlBN) and non-symmetric (ZMO-ZnO and AlBN-AlN) stacks, where the thickness ratio of the constituent layers are varied to identify the minimal thickness of alloyed material that is needed to switch the entire stack. Finally, we will propose an interface-based domain wall nucleation hypothesis to explain these observations.

CONTROL ID: 4000712
CURRENT SYMPOSIUM: S16 - New Frontiers and Science in Additive Manufacturing of Ceramic Materials
CURRENT SESSION: New approaches for AM processes of ceramics and ceramic-matrix composites
TITLE: Influence of humidity and thermal aging on the flow and packing behaviour of powders for additive manufacturing
PRESENTATION TYPE: Invited (by Invitation Only)
PRESENTER (FIRST NAME ONLY): Rafael
PRESENTER (LAST NAME ONLY): Kleba-Ehrhardt
PRESENTER (INSTITUTION ONLY): Technische Universitaet Berlin
PRESENTER (EMAIL ONLY): r.kleba-ehrhardt@ceramics.tu-berlin.de
PRESENTER (COUNTRY ONLY): Germany
DATE/TIME SUBMITTED: October 10, 2023, 06:45 AM
ABSTRACT BODY:
Abstract Body:
Powder-based additive manufacturing (AM), such as laser powder bed fusion (L-PBF) or binder jetting (BJ), have high demands on the properties of their feedstocks. The properties of the powder feedstocks largely determine the process reliability and reproducibility of additive manufacturing processes. While established measurement methods exist for the physical properties of powder particles (such as density, size and shape), there is currently no scientific consensus on characterizing the behavior of powders for and in AM applications. In particular, there are currently few detailed studies comparing the many different flowability measurement methods and discussing their specific suitability for additive manufacturing. In this work, five AM powder materials (metallic and ceramic) in the as-received condition, as well as modified powders, were investigated using a wide range of advanced powder flow characterization techniques (flow time, the Hausner-factor, the angle of repose, the angle of repose, the Scott bulk density, the powder rheology, the avalanche angle and the ring shear cell test were carried out).

CONTROL ID: 4001084

CURRENT SYMPOSIUM: S5 - Controlling Grain Boundary Structure, Chemistry, and their Network as a Function of Mtl Processing

CURRENT SESSION: Nanocrystalline ceramics

TITLE: $CuCr_2O_4$ Particle Growth and Evolution Across Sol-gel Routes and Calcination Profiles

PRESENTATION TYPE: Contributed (Oral)

PRESENTER (FIRST NAME ONLY): Ivar

PRESENTER (LAST NAME ONLY): Reimanis

PRESENTER (INSTITUTION ONLY): Colorado School of Mines

PRESENTER (EMAIL ONLY): reimanis@mines.edu

PRESENTER (COUNTRY ONLY): United States

DATE/TIME SUBMITTED: October 10, 2023, 07:53 PM

ABSTRACT BODY:

Abstract Body: $CuCr_2O_4$ spinel is a candidate coating material for central receivers in concentrating solar power to protect structural alloys against high temperature oxidation and related degradation. Coating performance and microstructure of dip-coated and sintered coatings is dictated by the initial particle size of the $CuCr_2O_4$ and sintering temperature, but can be compromised by particle agglomeration. In this study, sub-micron particles were synthesized through the Pechini and modified Pechini sol-gel methods. Phase composition was confirmed via X-ray diffraction. Particle growth during calcination of the nanoparticles at different temperatures (650°C, 750°C, 850°C) and times (between 1 h and 24 h) was measured via laser diffraction and scanning electron microscopy. The modified Pechini method displayed evidence of smaller particle sizes and greater agglomeration. The kinetics of particle growth observed are consistent with a diffusion limited inhibited grain growth model.

CONTROL ID: 4001090
CURRENT SYMPOSIUM: S5 - Controlling Grain Boundary Structure, Chemistry, and their Network as a Function of Mtl Processing
CURRENT SESSION: Microstructure evolution
TITLE: Microstructure of Solid Solutions between CeO_2 and Y_2O_3
PRESENTATION TYPE: Contributed (Oral)
PRESENTER (FIRST NAME ONLY): Alex
PRESENTER (LAST NAME ONLY): Hopkins
PRESENTER (INSTITUTION ONLY): Colorado School of Mines
PRESENTER (EMAIL ONLY): ahopkins1@mines.edu
PRESENTER (COUNTRY ONLY): United States
DATE/TIME SUBMITTED: October 10, 2023, 08:07 PM
ABSTRACT BODY:
Abstract Body: Yttrium oxide is of interest in missile window and dome applications which have transparency needs in the infrared and visible spectrum. There is a desire to improve the toughness of transparent ceramics via particle reinforcement, but without sacrificing high transparency. Precipitation of nanocrystallites from oxide solid solutions offer a potential method to form tough and transparent microstructures. The present talk describes experiments directed at controlling precipitation in solid solutions between CeO_2 and Y_2O_3. Samples of Y_2O_3 with 0, 15, 25 and 30 % CeO_2 were sintered for different times and temperatures in air and in vacuum. The microstructures are evaluated with x-ray diffraction, scanning electron microscopy, and transmission electron microscopy and are discussed here in relation to the processing conditions.

CONTROL ID: 4001094
CURRENT SYMPOSIUM: S15 - Electronics Materials for 5G/6G Applications
CURRENT SESSION: 5G/6G Metrology
TITLE: Tutorial: A review of some on-wafer permittivity measurement techniques
PRESENTATION TYPE: Contributed (Oral)
PRESENTER (FIRST NAME ONLY): Nate
PRESENTER (LAST NAME ONLY): Orloff
PRESENTER (INSTITUTION ONLY): NIST
PRESENTER (EMAIL ONLY): nathan.orloff@nist.gov
PRESENTER (COUNTRY ONLY): United States
DATE/TIME SUBMITTED: October 10, 2023, 08:24 PM
ABSTRACT BODY:
Abstract Body: In this talk, I will provide a tutorial reviewing the different devices people use to measure permittivity, the electrode design, and some trade-offs between different techniques. The goal of this talk to have a conversation about how the devices work and how you can incorporate them into your experimental designs. Having introduced the different types of devices, I will walk through how we would measure and calibrate the different devices. Then, I will walk through the analysis approach and data extraction. I will go through the procedure step-by-step and provide examples of ways to check your work as you go. I will wrap up by reviewing some simulation tools for each type of device and how we use them.

CONTROL ID: 4001096
CURRENT SYMPOSIUM: S15 - Electronics Materials for 5G/6G Applications
CURRENT SESSION: 5G/6G Devices
TITLE: Industry panel on 5G and 6G: Materials, Devices and Measurements
PRESENTATION TYPE: Invited (by Invitation Only)
PRESENTER (FIRST NAME ONLY): Nate
PRESENTER (LAST NAME ONLY): Orloff
PRESENTER (INSTITUTION ONLY): NIST
PRESENTER (EMAIL ONLY): nathan.orloff@nist.gov
PRESENTER (COUNTRY ONLY): United States
DATE/TIME SUBMITTED: October 10, 2023, 08:27 PM
ABSTRACT BODY:
Abstract Body: This panel will host members from companies developing new game changing technologies. Topics will include tunable materials, switches, acoustic materials, high-k dielectric for filters, and more. Please submit your questions ahead of time to the session chair.

CONTROL ID: 4001098
CURRENT SYMPOSIUM: S15 - Electronics Materials for 5G/6G Applications
CURRENT SESSION: Materials growth and design
TITLE: Introduction to session S15 – Electronics Materials for 5G/6G Applications
PRESENTATION TYPE: Contributed (Oral)
PRESENTER (FIRST NAME ONLY): Nate
PRESENTER (LAST NAME ONLY): Orloff
PRESENTER (INSTITUTION ONLY): NIST
PRESENTER (EMAIL ONLY): nathan.orloff@nist.gov
PRESENTER (COUNTRY ONLY): United States
DATE/TIME SUBMITTED: October 10, 2023, 08:30 PM
ABSTRACT BODY:
Abstract Body: Policy makers are turning their attention to new industry roadmaps for 6G technology. While the Federal Communications Administration has not auctioned any 6G bands, new roadmaps by the Alliance for Telecommunications Industry Solutions, the International Telecommunications Union and others set the operating frequency range at above 100 GHz, nearly 5 times higher than the millimeter-wave transceivers in most commercial handsets. At the same time, many 5G vendors are struggling to meet the measurement demands at 5G millimeter wave. In this talk, I will review current 6G roadmap documents and the unmet needs at 5G millimeter wave bands. I will discuss the organization of the session, the industry panel, and opportunities to interact during session S15 – Electronics Materials for 5G/6G Applications.

CONTROL ID: 4001100

CURRENT SYMPOSIUM: S3 - Advanced Electronic Materials: Processing and Structures and Properties and Applications

CURRENT SESSION: Advanced electronic materials, including ferroelectric, piezoelectric, dielectric, electrostrictive, electrocaloric and pyroelectric materials

TITLE: Improving equivalent circuit models of interdigitated capacitors

PRESENTATION TYPE: Invited (by Invitation Only)

PRESENTER (FIRST NAME ONLY): Nicholas

PRESENTER (LAST NAME ONLY): Jungwirth

PRESENTER (INSTITUTION ONLY): National Institute of Standards and Technology

PRESENTER (EMAIL ONLY): nrjungwirth@gmail.com

PRESENTER (COUNTRY ONLY): United States

DATE/TIME SUBMITTED: October 10, 2023, 08:52 PM

ABSTRACT BODY:

Abstract Body: Interdigitated capacitors are nearly ubiquitous circuits found in everything from telecommunications components to superconducting quantum computers. When the guided wavelength becomes comparable to the length of the interdigitation in the capacitor, common lumped-element models fail to describe how the impedance changes as a function of frequency. This failure limits the frequency range where one can accurately predict the performance of interdigitated capacitors and extract material properties. Here, we develop a multi-mode distributed theory for interdigitated capacitors and quantify the trade-offs between a lumped-element and distributed approach. To test our model, we measured interdigitated capacitors fabricated on $(LaAlO_3)_{0.3}(Sr_2TaAlO_6)_{0.7}$ substrate and interdigitated capacitors with same geometry fabricated on a $BaSrTiO_3$ thin film.

CONTROL ID: 4001105
CURRENT SYMPOSIUM: S12 - Magnetic and Superconducting Materials: From Basic Science to Applications
CURRENT SESSION: Superconducting and related magnetic materials
TITLE: Ultrafast Optically Pulse Triggered Microwave/Terahertz Emission from an Array of Charged Superconducting Rings
PRESENTATION TYPE: Contributed (Oral)
PRESENTER (FIRST NAME ONLY): Thomas
PRESENTER (LAST NAME ONLY): Bullard
PRESENTER (INSTITUTION ONLY): Wright Patterson AFB
PRESENTER (EMAIL ONLY): tbullard@vt.edu
PRESENTER (COUNTRY ONLY): United States
DATE/TIME SUBMITTED: October 10, 2023, 09:11 PM
ABSTRACT BODY:
Abstract Body: It is well established that current carrying superconducting materials will emit microwave/terahertz radiation when illuminated with a femtosecond infrared laser pulse. In this investigation we examine the electromagnetic emission from an inductively charged superconducting thin film ring. We believe this configuration lends itself to a simple compact microwave/terahertz emitter device as the ring simultaneously plays the part of an antenna, waveguide, and power supply. We find that the emitted microwave radiation of this system displays frequency dependence proportional to the size of the ring, a well-defined polarization direction, and a radiation pattern similar to that of an electrically large resonant loop antenna. With this knowledge we are able to construct an array of superconducting ring emitters. The resulting radiation pattern measurements indicated increased directionality as the number of elements increases indicating coherent emission. This suggests the possibility of a unique coherent pulsed microwave/terahertz source.

CONTROL ID: 4001601
CURRENT SYMPOSIUM: S15 - Electronics Materials for 5G/6G Applications
CURRENT SESSION: 5G/6G Metrology
TITLE: Comparing resonant cavity and on-wafer measurements of the complex permittivity of fused silica to 110 GHz
PRESENTATION TYPE: Contributed (Oral)
PRESENTER (FIRST NAME ONLY): Lucas
PRESENTER (LAST NAME ONLY): Enright
PRESENTER (INSTITUTION ONLY): National Institute of Standards and Technology
PRESENTER (EMAIL ONLY): lucas.enright@nist.gov
PRESENTER (COUNTRY ONLY): United States
DATE/TIME SUBMITTED: October 11, 2023, 05:07 PM
ABSTRACT BODY:
Abstract Body: Researchers rely on resonant cavity and on-wafer measurements of complex permittivity for circuit design and simulations. Resonant cavity measurements are more accurate but can only measure discrete (often single) frequencies. On-wafer measurements are less accurate but offer broadband measurements over huge frequency ranges. Previously, we collaborated with the International Electronics Manufacturing Initiative (iNEMI) and compared measurements from different resonant cavities in a series of round robin experiments. Those round robin experiments did not include any on-wafer measurements. Here, we compare resonant cavity and on-wafer complex permittivity measurements of the same fused silica wafer. We quantify the uncertainty of each method and the agreement between methods. Careful wafer design lets us cut coupons for resonant cavity measurements and chips for on-wafer measurements from a single wafer. Comparing and correlating these measurements gives us broadband complex permittivity (on-wafer) anchored with much smaller uncertainties at discrete frequencies (resonant cavities). These comparisons provide us with insight into frequency-dependent material behavior in fused silica as well as measurement capabilities and limitations.

CONTROL ID: 4001994

CURRENT SYMPOSIUM: S1: Characterization of Structure-Property Relationships in Functional Ceramics

CURRENT SESSION: Advances in connecting local and global structure to properties

TITLE: Phase formation Anomalies in Me^{2+}-Aluminates

PRESENTATION TYPE: Contributed (Oral)

PRESENTER (FIRST NAME ONLY): Mehmet

PRESENTER (LAST NAME ONLY): Gulgun

PRESENTER (INSTITUTION ONLY): Sabanci University

PRESENTER (EMAIL ONLY): mehmet.gulgun@sabanciuniv.edu

PRESENTER (COUNTRY ONLY): Turkey

DATE/TIME SUBMITTED: October 12, 2023, 10:27 AM

ABSTRACT BODY:

Abstract Body: Aluminum-oxygen or silicon-oxygen polyhedra form complex skeletons in many aluminate or silicate minerals. They usually incorporate other cations for charge compensation. Aluminate crystallography could be further complicated by that they can form either tetrahedral $(AlO_4)^{5-}$) or octahedral $((AlO_6)^{9-}$) polygons with oxygen. Many phase formation anomalies have been reported in literature in strontium and calcium aluminates. Dominating arguments for such observations were either the difficulty in diffusion of a large cation like Ca^{2+} or Sr^{2+} or complexity of polyhedral arrangements. Our recent observation during the early stages of precipitation in Ca-doped a-Al_2O_3 ceramics indicated that the first precipitate to form was $CaAl_2O_4$ and not $CaAl_{12}O_{19}$ that the equilibrium phase diagram would have suggested. Formation of a Ca-rich aluminate in alumina matrix speaks against the Ca-ion diffusion difficulty. We followed the incipient precipitation behavior in Sr- and Ca-aluminate system when the cations were mixed homogeneously at a molecular level, like in organic precursors using x-ray diffraction and electron imaging and spectroscopy. In contrast to this "diffusionless" scenario, we also studied the interface reactions in reaction couples made up of sapphire substrate and several organic precursor films with different Me^{2+} cation/Al^{3+} ratios deposited on the substrate using high resolution electron microscopy and spectroscopy.

CONTROL ID: 4002816
CURRENT SYMPOSIUM: S13 - Frontiers in Ferroic oxides: Synthesis, Properties, and Applications
CURRENT SESSION: Advanced synthesis and characterization techniques for ferroic oxides
TITLE: Direct Nanovolumetric Investigation of Domain Walls and Grain Boundaries in $ErMnO_3$
PRESENTATION TYPE: Contributed (Oral)
PRESENTER (FIRST NAME ONLY): Bryan
PRESENTER (LAST NAME ONLY): Huey
PRESENTER (INSTITUTION ONLY): University of Connecticut
PRESENTER (EMAIL ONLY): bryan.huey@uconn.edu
PRESENTER (COUNTRY ONLY): United States
DATE/TIME SUBMITTED: October 13, 2023, 03:58 PM
ABSTRACT BODY:
Abstract Body: Local properties are carefully investigated at domain boundaries and topological defects for manganate crystals and thin films. Leveraging Tomographic Atomic Force Microscopy especially with $ErMnO_3$, domain densities and orientations as a function of proximity from grain boundaries are investigated with nanoscale volumetric resolution. 3-dimensional localized curvature along domain walls, approaching domain intersections, and at grain boundaries are also uniquely resolved and statistically analyzed. Domain intersections exhibiting local chirality are directly observed as well. Such insights are important for advanced modelling, enhanced processing, and ultimately optimized applications of ferroic oxides.

CONTROL ID: 4004941
CURRENT SYMPOSIUM: S2 - Advanced Microelectronics
CURRENT SESSION: Flexible electronics
TITLE: Printed haptic module for enabling autonomous sensing in agricultural manipulators
PRESENTATION TYPE: Contributed (Oral)
PRESENTER (FIRST NAME ONLY): Prabha
PRESENTER (LAST NAME ONLY): Sundaravadivel
PRESENTER (INSTITUTION ONLY): The University of Texas at Tyler
PRESENTER (EMAIL ONLY): psundaravadivel@uttyler.edu
PRESENTER (COUNTRY ONLY): United States
DATE/TIME SUBMITTED: October 16, 2023, 11:30 AM
ABSTRACT BODY:
Abstract Body: This paper introduces the integration of printed haptic modules tailored for autonomous sensing in agricultural manipulators, revolutionizing modern farming practices. We aim to seamlessly incorporate these innovative modular tactile sensors to provide real-time data, enabling adaptive control of pressure specific to each application. This module offers a precise means of assessing crop health while minimizing the need for herbicides, thus presenting a promising solution for the agricultural sector. To realize this concept, we propose 3D printing the haptic module via inkjet technology on stretchable surfaces, which can be conveniently added to existing manipulators. We have explored the proposed haptic module on different stretchable surfaces including PLA and resin with different shore densities. Our haptic sensor's performance will be characterized based on pressure or vibration amplitude, haptic signal frequency, and feedback duration. Moreover, our module is designed to ensure a consistent distribution of haptic feedback across various contact points on a device. This research holds great potential to enhance agricultural practices, combining cutting-edge technology with sustainable farming techniques.

CONTROL ID: 4005456
CURRENT SYMPOSIUM: S17 - Electronic and Ionic Materials in Battery Energy Storage
CURRENT SESSION: Electrodes (Anodes and Cathodes)
TITLE: Structural and electrochemical properties of $Li_4Ti_5O_{12}$ epitaxial thin-film electrodes grown by pulsed laser deposition
PRESENTATION TYPE: Contributed (Oral)
PRESENTER (FIRST NAME ONLY): Jan
PRESENTER (LAST NAME ONLY): Zuntar
PRESENTER (INSTITUTION ONLY): Jozef Stefan Institute
PRESENTER (EMAIL ONLY): jan.zuntar@ijs.si
PRESENTER (COUNTRY ONLY): Slovenia
DATE/TIME SUBMITTED: October 16, 2023, 04:24 PM
ABSTRACT BODY:
Abstract Body: Owing to its high electrochemical stability and low volume change during (de)lithiation, $Li_4Ti_5O_{12}$ (LTO) is often utilized as the anode material in all-solid-state microbatteries. However, the deposition of fully epitaxial LTO thin film electrodes on Nb-doped $SrTiO_3$ (Nb:STO) single crystal substrates remains challenging due to the frequently reported occurrence of TiO_2 and Li_2TiO_3 secondary phases. Introducing conductive interlayers onto Nb:STO substrates may provide a more selectively favorable environment for epitaxial LTO growth for certain substrate orientations without negatively affecting the electrochemical performance.
In this work, LTO thin films were grown by pulsed laser deposition on Nb:STO single crystal substrates with (100), (110) and (111) orientations. Observation of RHEED patterns during thin-film growth indicated different LTO growth modes for the selected orientations, with (111) LTO exhibiting 2D-type and (100) LTO exhibiting 3D-type growth. XRD analysis of prepared LTO thin films confirmed the successful synthesis of epitaxial LTO for all three substrate orientations. A notable reduction of secondary phases for (100) and (110) orientations was achieved by the introduction of $SrRuO_3$ and La-doped $BaSnO_3$ layers, respectively.

CONTROL ID: 4005466
CURRENT SYMPOSIUM: S17 - Electronic and Ionic Materials in Battery Energy Storage
CURRENT SESSION: Electrodes (Anodes and Cathodes)
TITLE: Investigating structural changes in epitaxial $LiNi_{0.33}Mn_{0.33}Co_{0.33}O_2$ thin films by high-resolution X-ray diffraction
PRESENTATION TYPE: Contributed (Oral)

PRESENTER (FIRST NAME ONLY): Blaz
PRESENTER (LAST NAME ONLY): Jaklic
PRESENTER (INSTITUTION ONLY): Institute Josef Stefan
PRESENTER (EMAIL ONLY): blaz.jaklic@ijs.si
PRESENTER (COUNTRY ONLY): Slovenia

DATE/TIME SUBMITTED: October 16, 2023, 04:38 PM

ABSTRACT BODY:

Abstract Body: $LiNi_{0.33}Mn_{0.33}Co_{0.33}O_2$ (NMC 111) thin film cathodes were prepared on (001) out-of-plane oriented $SrRuO_3$/Nb:$SrTiO_3$ substrates using pulsed laser deposition. Structural properties of pristine NMC 111 thin films were characterized by high-resolution X-ray diffraction (HRXRD). HRXRD analysis revealed epitaxial growth of NMC 111 thin films in (104) out-of-plane orientation with 4-fold symmetry and 45° in-plane tilt with respect to the substrate. Accurate measurements of (104) and (107) planes in reciprocal space were used to calculate lattice parameters and the volume of the NMC 111 unit cell, while twin boundaries, observed by scanning transmission electron microscopy, proved to be the reason for the high mosaicity in NMC 111 thin films. Additionally, ex-situ HRXRD was performed on cycled NMC 111 thin film in delithiated state (charged to 4.2 V vs. Li/Li^+) and lithiated state (discharged to 3.0 V vs. Li/Li^+). Delithiation resulted in negative volume change and induced twinning of the crystal, indicating instability of the thin film cathode in the charged state. After discharging, the structure of NMC 111 is reversibly preserved, as the unit cell volume and measure of mosaicity are comparable to the film in the pristine state.

CONTROL ID: 4005491

CURRENT SYMPOSIUM: S4 - Complex Oxide Thin Films and Heterostructures: From Synthesis to Strain/Interface-engineered Emergent Properties

CURRENT SESSION: Controlled synthesis of epitaxial thin films and nanocomposites, including heterostructures, superlattices, and vertically aligned nanocomposites for functional properties

TITLE: $SrTiO_3$ Passivation of Silicon Photocathode for Solar Water Splitting

PRESENTATION TYPE: Contributed (Oral)

PRESENTER (FIRST NAME ONLY): Matjaz

PRESENTER (LAST NAME ONLY): Spreitzer

PRESENTER (INSTITUTION ONLY): Jozef Stefan Institute

PRESENTER (EMAIL ONLY): matjaz.spreitzer@ijs.si

PRESENTER (COUNTRY ONLY): Slovenia

DATE/TIME SUBMITTED: October 16, 2023, 04:50 PM

ABSTRACT BODY:

Abstract Body: Development of a robust photocathode using low-cost and high-performing materials, e.g., p-Si, to produce clean fuel hydrogen has remained challenging since the semiconductor substrate is easily susceptible to (photo)corrosion under photoelectrochemical (PEC) operational conditions. In the present work, we utilized pulsed laser deposition (PLD) to prepare a high-quality SrTiO3 (STO) layer to passivate the p-Si substrate using a buffer layer of reduced graphene oxide (rGO). Specifically, a very thin (3.9 nm ~10 unit cells) STO layer epitaxially overgrown on rGO-buffered Si showed the highest onset potential (0.326 V vs RHE) in comparison to the counterparts with thicker and/or nonepitaxial STO. The photovoltage, flat-band potential, and electrochemical impedance spectroscopy measurements revealed that the epitaxial photocathode was more beneficial for charge separation, charge transfer, and targeted redox reaction than the nonepitaxial one. The STO/rGO/Si with a smooth and highly epitaxial STO layer outperforming the directly contacted STO/Si with a textured and polycrystalline STO layer showed the importance of having a well-defined passivation layer. In addition, the numerous pinholes formed in the directly contacted STO/Si led to the rapid degradation of the photocathode during the PEC measurements.

CONTROL ID: 4009482
CURRENT SYMPOSIUM: S2 - Advanced Microelectronics
CURRENT SESSION: Dielectrics and ferroelectrics for device applications
TITLE: Manipulating Magneto-Elasto-Electric Coupling in Multiferroic Nanocomposites
PRESENTATION TYPE: Invited (by Invitation Only)
PRESENTER (FIRST NAME ONLY): Ruyan
PRESENTER (LAST NAME ONLY): Guo
PRESENTER (INSTITUTION ONLY): The University of Texas at San Antonio
PRESENTER (EMAIL ONLY): ruyan.guo@utsa.edu
PRESENTER (COUNTRY ONLY): United States
DATE/TIME SUBMITTED: October 18, 2023, 05:19 PM
ABSTRACT BODY:
Abstract Body: The presentation explores the design and fabrication of electronic and photonic tunable devices utilizing ferroelectric and multiferroic composite materials, highlighting magneto-elasto-electric coupling in multiferroic nanocomposites. Examples that demonstrate transducer effect from low-frequency piezoelectric excitation to RF amplitude modulation in layered composite structures and directional wave propagation in core-shell nanocomposites resulting in amplitude and phase modulations, are reviewed. The talk discusses current research on controlling electromagnetic wave propagation through interactions with surface and interface charges, along with advancements in hybrid ink jet fabrication for layered ferroic and patterned core-shell multiferroic composite materials.

CONTROL ID: 4011749
CURRENT SYMPOSIUM: S14 - Emerging Semiconductor Materials and Interfaces
CURRENT SESSION: WBG and UWBG materials - synthesis, characterization, theory
TITLE: Gallium Oxide and its Heterogenous Integration
PRESENTATION TYPE: Invited (by Invitation Only)
PRESENTER (FIRST NAME ONLY): James
PRESENTER (LAST NAME ONLY): Pomeroy
PRESENTER (INSTITUTION ONLY): University of Bristol
PRESENTER (EMAIL ONLY): James.Pomeroy@bristol.ac.uk
PRESENTER (COUNTRY ONLY): United Kingdom
DATE/TIME SUBMITTED: October 20, 2023, 02:36 PM
ABSTRACT BODY:
Abstract Body: Gallium Oxide with its low thermal conductivity and challenge to enable workable p-type doping benefits from integration with other materials. This includes for example integration with diamond and also SiC. Their increased thermal conductivity will benefit heat extraction from the devices. Latest advances in device concepts that use integrated materials, also advances in integration by growth will be discussed; we also highlight techniques to be able to measure the thermal implications of this integration e.g. on device temperature with very high spatial resolution.

CONTROL ID: 4011768
CURRENT SYMPOSIUM: S12 - Magnetic and Superconducting Materials: From Basic Science to Applications
CURRENT SESSION: Superconducting and related magnetic materials
TITLE: Distance-Dependent Evolution of Electronic States in Kagome-Honeycomb Lateral Heterostructures in FeSn
PRESENTATION TYPE: Contributed (Oral)
PRESENTER (FIRST NAME ONLY): Seoung-Hun
PRESENTER (LAST NAME ONLY): Kang
PRESENTER (INSTITUTION ONLY): Oak Ridge National Laboratory
PRESENTER (EMAIL ONLY): kangs@ornl.gov
PRESENTER (COUNTRY ONLY): United States
DATE/TIME SUBMITTED: October 20, 2023, 03:31 PM
ABSTRACT BODY:
Abstract Body: Lateral heterostructures offer spatially resolved property control, enabling the integration of dissimilar materials, and promoting novel phenomena not typically observed in vertical heterostructures. Using the first-principles density functional theory approach, we investigate the stability and electronic structures of lateral heterointerfaces in FeSn containing Kagome and honeycomb layers. We provide a comprehensive understanding of the bonding motif connecting the Fe3Sn-terminated Kagome and Sn2-terminated honeycomb surfaces. More importantly, we reveal a unique distance-dependent evolution of the electronic states in the vicinity of the heterointerfaces. This evolution is significantly influenced by the orbital character of the flat bands. Our findings also confirm by the scanning tunnelling microscopy/spectroscopy on sample made with molecular beam epitaxy technique. Our findings offer a new perspective for modulating the electronic properties of the Kagome lattice, which should be beneficial for the development of future quantum devices.

CONTROL ID: 4012670
CURRENT SYMPOSIUM: S14 - Emerging Semiconductor Materials and Interfaces
CURRENT SESSION: WBG and UWBG materials - synthesis, characterization, theory
TITLE: Gallium Oxide Wafer Technology for Power Device Applications
PRESENTATION TYPE: Invited (by Invitation Only)
PRESENTER (FIRST NAME ONLY): Akito
PRESENTER (LAST NAME ONLY): Kuramata
PRESENTER (INSTITUTION ONLY): Novel Crystal Technology, Inc.
PRESENTER (EMAIL ONLY): kuramata@novelcrystal.co.jp
PRESENTER (COUNTRY ONLY): Japan
DATE/TIME SUBMITTED: October 23, 2023, 10:50 AM
ABSTRACT BODY:
Abstract Body: Ga_2O_3 is expected to be used as a material for next generation power devices. Due to its large breakdown electric field, Ga_2O_3 is suitable for high voltage applications. Ga_2O_3 is a material that can be manufactured at low cost because it can be grown with high growth rates from melt and its hardness is not high. Currently, 100mm substrates manufactured using the edge-defined film-fed growth (EFG) method are commercially available. 100mm epitaxial wafers grown using the halide vapor phase epitaxy (HVPE) method have also been commercialized. Although Ga_2O_3 devices are not yet commercially available, didoes and transistors with ampere-class currents and breakdown voltages of over 1 kV have been demonstrated so far.

When considering material selection for power devices, the availability of native substrates is important. Ga_2O_3 meets this requirement. This presentation will report on wafer fabrication technology of Ga_2O_3. We report on two bulk growth techniques. One is the EFG, and another is the vertical Bridgman (VB) method. We also report on epitaxial growth technology using the HVPE method.

CONTROL ID: 4013740
CURRENT SYMPOSIUM: S14 - Emerging Semiconductor Materials and Interfaces
CURRENT SESSION: WBG and UWBG materials - synthesis, characterization, theory
TITLE: Homoepitaxial Metalorganic Vapor Phase Epitaxy Growth of (010) β-Ga_2O_3 Films For Lateral and Vertical Power Devices
PRESENTATION TYPE: Invited (by Invitation Only)
PRESENTER (FIRST NAME ONLY): Sriram
PRESENTER (LAST NAME ONLY): Krishnamoorthy
PRESENTER (INSTITUTION ONLY): University of California, Santa Barbara
PRESENTER (EMAIL ONLY): sriramkrishnamoorthy@ucsb.edu
PRESENTER (COUNTRY ONLY): United States
DATE/TIME SUBMITTED: October 24, 2023, 10:17 AM
ABSTRACT BODY:
Abstract Body: This work focuses on growth, doping, epitaxial stack engineering, and characterization of β-Ga_2O_3 thin films for lateral devices and thick epitaxial drift layers for vertical power devices. β-Ga_2O_3 is an emerging ultra-wide bandgap (UWBG) semiconductor material that offers projected size, weight, and power (SWaP) performance advantage over the incumbent power semiconductors for solid-state power switching applications due to its high breakdown field strength (~ 8 MV/cm). We demonstrate a hybrid low temperature - high temperature (LT-HT) buffer/channel stack growth using MOVPE with record carrier mobility values (range of 196–85 cm^2/Vs) over four orders of doping range (2×10^{16} – 1×10^{20} cm^{-3}). Record electron mobility of 110 cm^2/Vs is also demonstrated in delta-doped (2D) channels (n_s = 9.2×10^{12} cm^{-2}). The improvement in transport properties was achieved mainly by realizing pristine doped channels, eliminating undesired parasitic conduction paths, and minimizing carrier compensation. We will also discuss growth of thick epitaxial drift layers for vertical high voltage devices with high growth rates, low background carrier concentration and record drift layer mobility.

CONTROL ID: 4014559
CURRENT SYMPOSIUM: S15 - Electronics Materials for 5G/6G Applications
CURRENT SESSION: Materials growth and design
TITLE: An On-Chip Electric-Acoustic Force Balance for Particle Sorting
PRESENTATION TYPE: Contributed (Oral)
PRESENTER (FIRST NAME ONLY): Leya
PRESENTER (LAST NAME ONLY): Shaw
PRESENTER (INSTITUTION ONLY): University of Colorado, Boulder
PRESENTER (EMAIL ONLY): leya.shaw@colorado.edu
PRESENTER (COUNTRY ONLY): United States
DATE/TIME SUBMITTED: October 25, 2023, 10:40 AM
ABSTRACT BODY:
Abstract Body: Surface acoustic waves (SAW) in lab-on-a-chip microfluidics have demonstrated the potential for efficient processing of particles using acoustic focusing driven by microwave sources. This method of trapping serves as a general platform for particle manipulation, but further sorting of particles based on desirable electronic properties has been underexplored. If acoustic trapping forces could be paired with electrophoretic and dielectrophoretic forces, we could create a force balance for spatially separating and sorting different species of particles. In this work, we develop an on-chip electric-acoustic force balance targeted at recovering rare earth metals from granular electronic waste. The system is composed of SAW resonator chips that are integrated with microfluidics and are attached to a modular platform atop a printed circuit board. The chips contain two-port (transmission) SAW resonators on lithium niobate that launch Rayleigh waves in opposing directions across a 75 micron gap. Inside this gap, we placed a ground-signal (GS) microwave transmission line and microfluidic channel. The GS line allows us to apply DC and microwave fields that apply a force to particles trapped by the acoustic field, creating the force balance. We have demonstrated tunable acoustic trapping by varying the inputted microwave amplitude and phase.

CONTROL ID: 4014755
CURRENT SYMPOSIUM: S19 - Failure: The Greatest Teacher
CURRENT SESSION: None
TITLE: It's what you know for sure that just ain't so
PRESENTATION TYPE: Invited (by Invitation Only)
PRESENTER (FIRST NAME ONLY): Vladan
PRESENTER (LAST NAME ONLY): Stevanovic
PRESENTER (INSTITUTION ONLY): Colorado School of Mines
PRESENTER (EMAIL ONLY): vstevano@mines.edu
PRESENTER (COUNTRY ONLY): United States
DATE/TIME SUBMITTED: October 25, 2023, 02:34 PM
ABSTRACT BODY:
Abstract Body: It's what you know for sure that just ain't so

CONTROL ID: 4014759
CURRENT SYMPOSIUM: S19 - Failure: The Greatest Teacher
CURRENT SESSION: None
TITLE: Cook and look or think and drink: Solid state synthesis of powder, microstructures, researchers and careers
PRESENTATION TYPE: Invited (by Invitation Only)
PRESENTER (FIRST NAME ONLY): Wolfgang
PRESENTER (LAST NAME ONLY): Rheinheimer
PRESENTER (INSTITUTION ONLY): University of Stuttgart
PRESENTER (EMAIL ONLY): wolfgang.rheinheimer@ifkb.uni-stuttgart.de
PRESENTER (COUNTRY ONLY): Germany
DATE/TIME SUBMITTED: October 25, 2023, 02:39 PM
ABSTRACT BODY:
Abstract Body: Cook and look or think and drink: Solid state synthesis of powder, microstructures, researchers and careers.

CONTROL ID: 4015381
CURRENT SYMPOSIUM: S14 - Emerging Semiconductor Materials and Interfaces
CURRENT SESSION: Inorganic perovskites - synthesis, characterization, theory
TITLE: Metal halide perovskites: Emerging materials for light emission and spin manipulation
PRESENTATION TYPE: Invited (by Invitation Only)
PRESENTER (FIRST NAME ONLY): Young-Hoon
PRESENTER (LAST NAME ONLY): Kim
PRESENTER (INSTITUTION ONLY): Hanyang University
PRESENTER (EMAIL ONLY): younghoonkim@hanynang.ac.kr
PRESENTER (COUNTRY ONLY): Korea (the Republic of)
DATE/TIME SUBMITTED: October 26, 2023, 03:05 PM
ABSTRACT BODY:
Abstract Body: We report a strategy to produce spin-polarized carriers in metal halide perovskite (MHPs) and demonstrate a spin light-emitting diode (LED) that operates at room temperature and without magnetic fields or ferromagnetic contacts. We employ the chiral induced spin selectivity (CISS) layer which consists of oriented self-assembled small chiral molecules within a layered MHPs. The CISS layer is used to transport spin-polarized holes into an achiral colloidal perovskite nanocrystal with a spin-injection selectivity of up to 80%. In perovskite nanocrystal emitting layers, injected spin-polarized holes recombine with electrons which have proper spin direction and emit light with a matching polarization. Additionally, mixed-halide is utilized to facilitate the radiative recombination of spin-polarized excitons while suppressing spin-dephasing. The spin-LED achieves ± 2.6% circularly-polarized electroluminescence at room temperature. Thus, here we control of spin, charge, and light with only an applied electric field.

CONTROL ID: 4015807
CURRENT SYMPOSIUM: S5 - Controlling Grain Boundary Structure, Chemistry, and their Network as a Function of Mtl Processing
CURRENT SESSION: Microstructure evolution
TITLE: Low temperature SPS sintering of $SrTiO_3$: Taking advantage of novel processing techniques for grain boundary engineering in functional ceramics
PRESENTATION TYPE: Contributed (Oral)
PRESENTER (FIRST NAME ONLY): Marc
PRESENTER (LAST NAME ONLY): Zahler
PRESENTER (INSTITUTION ONLY): Forschungszentrum Juelich
PRESENTER (EMAIL ONLY): m.zahler@fz-juelich.de
PRESENTER (COUNTRY ONLY): Germany
DATE/TIME SUBMITTED: October 27, 2023, 01:11 PM
ABSTRACT BODY:
Abstract Body: Conduction processes across grain boundaries in ceramics are often hindered by electrostatic potential barriers at these grain boundaries; so-called space charge layers. For that reason, scientists try to understand mechanisms of space charge layer formation and find new ways how to suppress or control them. Spark plasma sintering (SPS) sintering and cold-sintering give the unique opportunity to densify ceramic bodies at temperatures lower than conventional sintering temperatures slowing down cationic diffusion significantly which may prevent the formation of space charge layers.

In this study $SrTiO_3$ was sintered using low temperature SPS to suppress the formation of space charge layers. Ceramic powders of different composition of $SrCO_3$, TiO_2 and $SrTiO_3$ were prepared and sintered at a variety of temperatures and pressures. Samples densities, microstructures and compositions were analyzed using Archimedes method, SEM and XRD. The conductivity of SPS sintered samples was analyzed using electrochemical impedance spectroscopy (EIS). EIS and STEM-EDS were used to investigate space charge layers and segregation behavior of cationic defects. All data were compared to conventionally sintered $SrTiO_3$ samples and discussed within the framework of defect chemistry and space charge layer models.

CONTROL ID: 4015815
CURRENT SYMPOSIUM: S12 - Magnetic and Superconducting Materials: From Basic Science to Applications
CURRENT SESSION: Other magnetic materials
TITLE: Microwave magnetodielectric materials and emerging applications
PRESENTATION TYPE: Invited (by Invitation Only)
PRESENTER (FIRST NAME ONLY): Yajie
PRESENTER (LAST NAME ONLY): Chen
PRESENTER (INSTITUTION ONLY): Rogers Corporation
PRESENTER (EMAIL ONLY): yajie.chen@rogerscorp.com
PRESENTER (COUNTRY ONLY): United States
DATE/TIME SUBMITTED: October 27, 2023, 01:23 PM
ABSTRACT BODY:
Abstract Body: With the dramatic evolution of modern communication technology, magnetodielectric materials have been increasingly drawing attention especially in RF and microwave communication fields. In this presentation, we will briefly review the history and challenges of the development of magneto-dielectric materials and introduce MAGTREX^{TM} materials, a novel class of impedance controlled magnetodielectric materials with one launched product and other grades under development at Rogers. Rogers is building out a platform of MAGTREX^{TM} materials that includes engineered polymer composites/laminates and single-phase ceramics. These materials display low magnetic loss, high permeability, and controlled impedance at operating frequencies ranging from 100 MHz to 9 GHz, which can be used in applications like antennas, inductors, and filters to be part of systems such as GSM, GNSS/GPS, Wifi, cellular, satellite communications, etc. Especially, these materials have demonstrated capabilities in enhancing performance of antennas at different bands or applications, compared to conventional dielectric substrates. The materials under development enable miniaturization, wide bandwidth and reconfigurable ability for highly reliable microwave devices. As a result, present work has been closing the gap between concept and practice, making what was once a wished for combination of material properties into a commercial reality.

CONTROL ID: 4039741
CURRENT SYMPOSIUM: S9 - Machine Learning and Automated Synthesis/Characterization for Novel Materials
CURRENT SESSION: Computer-assisted experiment automation
TITLE: Quantitative nano-ionic dynamics via functional AFM and Gaussian process optimization
PRESENTATION TYPE: Invited (by Invitation Only)
PRESENTER (FIRST NAME ONLY): Marti
PRESENTER (LAST NAME ONLY): Checa
PRESENTER (INSTITUTION ONLY): Oak Ridge National Lab
PRESENTER (EMAIL ONLY): checam@ornl.gov
PRESENTER (COUNTRY ONLY): United States
DATE/TIME SUBMITTED: November 30, 2023, 12:50 PM
ABSTRACT BODY:
Abstract Body: Unraveling local dynamic charge processes is vital for progress in diverse fields, from microelectronics to energy storage. This relies on the ability to map charge carrier motion across multiple length- and timescales and understanding how these processes interact with the inherent material heterogeneities. Towards addressing this challenge, we introduce high-speed Sparse Scanning Kelvin Probe Force Microscopy (SS-KPFM), which combines sparse spiral scanning and image reconstruction via Gaussian process optimization. SS-KPFM enables sub-second imaging rates (3.3 fps) of nanoscale charge dynamics, representing several orders of magnitude improvement over traditional KPFM methods. Bridging this improved spatiotemporal resolution with macroscale device measurements, we demonstrate the applicability of this approach by visualizing electrochemically mediated diffusion of mobile surface ions on a $LaAlO_3/SrTiO_3$ planar device, processes which are known to impact band-alignment and charge-transfer dynamics at such heterointerfaces. Additionally, we track oxygen vacancy diffusion generated under a biased tip, at the single grain level in polycrystalline TiO_2. Temperature-dependent measurements reveal a charge diffusion activation energy of 0.18 eV, in good agreement with previously reported values and backed up here by further modelling.

CONTROL ID: 4044109
CURRENT SYMPOSIUM: S17 - Electronic and Ionic Materials in Battery Energy Storage
CURRENT SESSION: Electrodes (Anodes and Cathodes)
TITLE: Rationally Engineering Interfaces to Improve Performances of Li Metal Batteries
PRESENTATION TYPE: Invited (by Invitation Only)
PRESENTER (FIRST NAME ONLY): Bin
PRESENTER (LAST NAME ONLY): Li
PRESENTER (INSTITUTION ONLY): Oak Ridge National Lab
PRESENTER (EMAIL ONLY): lib2@ornl.gov
PRESENTER (COUNTRY ONLY): United States
DATE/TIME SUBMITTED: December 04, 2023, 10:20 AM
ABSTRACT BODY:
Abstract Body: Lithium (Li) metal batteries have attracted world-wide attention due to its low density (0.534 g/cm^3), high theoretical capacity (3,860 mAh/g), and low electrochemical potential (-3.04 V vs. SHE), enabling their potential to double the cell-level energy of the state-of-the-art lithium-ion batteries. However, there are still key materials issues that inhibit lithium metal from being commercialized as an anode for rechargeable battery applications. The chemical and morphological (e.g., grain distribution, surface roughness, etc.) non-uniformity, coupled with the high reactivity of lithium drive heterogeneous solid-electrolyte interphase (SEI) formation, poor ionic flux distribution, active material consumption, and rapid dendrite growth. Currently, there are ex-situ and in-situ approaches to address the above issues. For example, one of approaches is to coat artificial SEI layers on Li metal anode surfaces. The component, distribution and structures of SEI layers should be designed and well controlled. The other approach is to exploit advanced electrolyte to in-situ form high-performance SEI layers. In this talk, we will present our recent work from the two aspects: a) novel closed-host bi-layer porous/dense artificial SEI layers were designed; b) the microstructures of localized high-concentration electrolyte were deeply understood.

CONTROL ID: 4044351
CURRENT SYMPOSIUM: S14 - Emerging Semiconductor Materials and Interfaces
CURRENT SESSION: Inorganic perovskites - synthesis, characterization, theory
TITLE: New Class Semiconducting Materials of Organic-Inorganic Metal Halides : Synthesis, Structure, and Properties
PRESENTATION TYPE: Invited (by Invitation Only)
PRESENTER (FIRST NAME ONLY): Hyunjung
PRESENTER (LAST NAME ONLY): Shin
PRESENTER (INSTITUTION ONLY): SungKyunKwan University
PRESENTER (EMAIL ONLY): hshin@skku.edu
PRESENTER (COUNTRY ONLY): Korea (the Republic of)
DATE/TIME SUBMITTED: December 04, 2023, 01:27 PM
ABSTRACT BODY:
Abstract Body: Organic-inorganic hybrid metal halides are now emerging as new class of semiconductors. It shows Perovskite crystal structure and is used as a main light absorber in photovoltaic cells, known as 'Perovskite Solar Cell (PSC)'. The notable examples are $MAPbI_3$ (methyl ammonium lead triiodides) and $FAPbI_3$ (formamidinium lead triiodides). The PSC now shows a remarkable power conversion efficiency (PCE) of 26.1%. These halides are truly unique semiconducting materials showing excellent optoelectronic properties, even their preparation methods are simple like spin coatings of precursor solution with a low temperature annealing. Unlike monocrystalline Si, and GaAs, these halides contain lot of defects, which are potentially non-radiative recombination centers. Defect tolerating in these halides is now under debating and thus defect engineering is the key to increase further the PCE. In this talk, the unique semiconducting properties studied are summarized and understanding of the materials' structure-properties relationship will be discussed.

CONTROL ID: 4057557

CURRENT SYMPOSIUM: S16 - New Frontiers and Science in Additive Manufacturing of Ceramic Materials

CURRENT SESSION: New approaches for AM processes of ceramics and ceramic-matrix composites

TITLE: Advancing Energy Solutions through Innovative Ceramic Manufacturing: Highlights from ARPA-E Projects on Magnetic Materials and High-Temperature Heat Exchangers

PRESENTATION TYPE: Invited (by Invitation Only)

PRESENTER (FIRST NAME ONLY): Philseok

PRESENTER (LAST NAME ONLY): Kim

PRESENTER (INSTITUTION ONLY): US Department of Energy

PRESENTER (EMAIL ONLY): phil.kim@hq.doe.gov

PRESENTER (COUNTRY ONLY): United States

DATE/TIME SUBMITTED: December 13, 2023, 01:46 PM

ABSTRACT BODY:

Abstract Body: This presentation introduces the Advanced Research Projects Agency – Energy (ARPA-E) under the U.S. Department of Energy and provides insights into ongoing projects focusing on ceramic additive manufacturing. Notable example projects include 1) leveraging 3D-printing techniques for the production of amorphous metal Soft Magnetic Composites (SMCs) to optimize costs, minimize material waste, and enhance design flexibility, potentially achieving magnetic cores with significantly reduced losses, 2) redefining manufacturing paradigm for electric machines by employing Composite Magnetic Powders (CMPs) and electrophoretic deposition of aluminum nitride insulators to increase the efficiency and reduce production costs significantly, 3) validating the feasibility of advanced ceramic-based 3D printing for ultra-compact, high-temperature, high-pressure heat exchangers with exceptional performance under extreme conditions, and 4) enabling 3D packaging solutions and thermal management of power electronics via in situ reactive 3D printing of ceramics. These endeavors collectively underscore the transformative impact of cutting-edge manufacturing techniques in various energy applications, spanning converters, storage devices, packaging, thermal systems, and electric vehicles.

CONTROL ID: 4072939
CURRENT SYMPOSIUM: S18 - Structure and function in high-entropy, entropy-stabilized, and related oxides
CURRENT SESSION: Characterization and quantification of structure and properties
TITLE: High-Entropy Pb-based Perovskite Local Structure Analysis through fitting of EXAFS data
PRESENTATION TYPE: Poster
PRESENTER (FIRST NAME ONLY): John
PRESENTER (LAST NAME ONLY): Barber
PRESENTER (INSTITUTION ONLY): Virginia Tech
PRESENTER (EMAIL ONLY): jpbarber38@vt.edu
PRESENTER (COUNTRY ONLY): United States
DATE/TIME SUBMITTED: January 16, 2024, 04:42 PM
ABSTRACT BODY:
Abstract Body: Numerous devices require small and efficient coolant systems. The electrocaloric effect (ECE) as a method for solid-state cooling provides an avenue to refrigeration. Very large ECE has been demonstrated in lead zirconate-titanate (PZT) thin films and in relaxor ferroelectrics, where polar nanoregions are typically responsible for the polarization required to achieve large ECE. Here, the high-entropy approach is used to synthesize PZT-based perovskites with a compositionally disordered B-site occupied by Hf, Zr, Ti, Nb, and either Fe or Sc ($Pb(HfZrTiNbX)_{0.2}O_3$, X=Fe, Sc). X-ray absorption spectroscopy (XAS) was conducted at the Advanced Photon Source, Argonne National Lab on bulk samples to characterize the local geometry of select elements including hafnium, zirconium, and niobium. This body of work demonstrates phase analysis of x-ray diffraction data and details the process used to fit and analyze the extended x-ray absorption fine structure (EXAFS) data. Analysis of the EXAFS data indicates varying local structure across B-site cations within the samples.

CONTROL ID: 4072952

CURRENT SYMPOSIUM: S15 - Electronics Materials for 5G/6G Applications

CURRENT SESSION: Materials growth and design

TITLE: Microwave Dielectric Properties and Loss Mechanism of $(Ca_{1+x}Sm_{1-x})(Al_{1-x}Ti_x)O_4$ Ceramics

PRESENTATION TYPE: Poster

PRESENTER (FIRST NAME ONLY): Mingyu

PRESENTER (LAST NAME ONLY): Kim

PRESENTER (INSTITUTION ONLY): Changwon National University

PRESENTER (EMAIL ONLY): rlaalsrb119@naver.com

PRESENTER (COUNTRY ONLY): Korea (the Republic of)

DATE/TIME SUBMITTED: January 15, 2024, 10:59 PM

ABSTRACT BODY:

Abstract Body: According to the interest in 5G+/6G wireless communication technology getting higher, their basic technologies have also been starting to attract attention. With increasing application frequency range, microwave dielectrics, which have low-loss and middle-permittivity (k~20) dielectric properties that are adaptable to RF filters, are required especially in wireless communication base stations. It has been known that K_2NiF_4-type structured dielectrics layered by ABO_3 perovskite structure could be a candidate since they have middle-permittivity, low-loss, and temperature-stable properties. In this study, we considered that $CaSmAlO_4$-based dielectrics would be a promising material by controlling their microstructure and crystal structure. $(Ca_{1+x}Sm_{1-x})(Al_{1-x}Ti_x)O_4$ (x = 0.1, 0.15, 0.2) ceramics were synthesized at 1300 °C by using a conventional solid-state reaction method, and the sintering was performed at 1400 - 1600 °C. The microwave dielectric properties were analyzed in terms of microstructure and lattice distortion. Consequently, we have confirmed that microstructure and lattice distortion are correlated with variations in permittivity (ε_r) and dielectric loss. Specifically, it is considered that dielectric loss, which can be expressed in the quality factor (Q_f), is mainly affected by lattice distortion rather than the microstructure of the ceramics.

CONTROL ID: 4075634
CURRENT SYMPOSIUM: S14 - Emerging Semiconductor Materials and Interfaces
CURRENT SESSION: Inorganic perovskites - synthesis, characterization, theory
TITLE: Lead-free halide perovskite light-emitting diodes using vacuum thermal evaporation
PRESENTATION TYPE: Poster
PRESENTER (FIRST NAME ONLY): Do Young
PRESENTER (LAST NAME ONLY): Kim
PRESENTER (INSTITUTION ONLY): Oklahoma State University
PRESENTER (EMAIL ONLY): doyoung.kim@okstate.edu
PRESENTER (COUNTRY ONLY): United States
DATE/TIME SUBMITTED: January 18, 2024, 07:57 PM
ABSTRACT BODY:
Abstract Body: Perovskite based visible light emitting diodes (LEDs) with blue, green, and red emissions have successfully been demonstrated with high external quantum efficiencies (EQEs) and bright luminance for next-generation display and solid-state lighting applications. Despite the high performances of perovskite-based visible LEDs, however, their applications will be limited due to the toxicity of lead-based halide perovskites. While there have currently been many reports about lead-free metal halide emitters to solve this toxicity issue, most lead-free halide LEDs show still significantly low performance, resulting in below 1 %.
In this work, lead-free halide LEDs with $CsCu_2I_3$ as a lead-free halide emitter are fabricated with a strong yellow emission at a wavelength of 568 nm. To fabricate high-efficiency lead-free halide LEDs, $CsCu_2I_3$ thin film is deposited by a vacuum thermal evaporation process instead of a typical solution-based halide emitter thin film process. Our $CsCu_2I_3$-based lead-free halide LED finally shows a significantly high EQE of more than 7 %, which is the highest efficiency in the world as the lead-free halide LEDs. Various device structures have systematically been studied for maximizing LED performance in this study. In this talk, we will present in detail how we can successfully achieve such high-efficiency lead-free halide LEDs.

CONTROL ID: 4075646
CURRENT SYMPOSIUM: S1: Characterization of Structure-Property Relationships in Functional Ceramics
CURRENT SESSION: Advances in scattering, imaging, and analytical techniques
TITLE: Water splitting at the phase transition of ferroelectric oxide single crystals
PRESENTATION TYPE: Poster
PRESENTER (FIRST NAME ONLY): Neus
PRESENTER (LAST NAME ONLY): Domingo Marimon
PRESENTER (INSTITUTION ONLY): Oak Ridge National Lab
PRESENTER (EMAIL ONLY): ndomingo@ornl.gov
PRESENTER (COUNTRY ONLY): United States
DATE/TIME SUBMITTED: January 18, 2024, 09:34 PM
ABSTRACT BODY:
Abstract Body: Ferroelectric surfaces are amazing playgrounds for reduction and oxidation reactions since they permanently accumulate surface charges and fields and show specific electrochemical reactivity as a function of the ferroelectric polarization [1,2,3]. Among surface electrochemical reactions, water splitting is naturally observed on FE oxide surfaces and hence hints at their application not only for water decomposition (i.e. H2 production) but also as new platforms for catalysis. This feature opens new opportunities such as ferrocatalysis, that is, to exploit ferroelectric surfaces in catalysis for water splitting since ferroelectric polarization becomes a switch to adjust surface catalytic properties.[4]

Here, I will present a study of the water splitting reactions on ferroelectric surfaces as a function of polarization, obtained from the combination of near ambient pressure XPS analysis and Scanning Probe Microscopy techniques (essentially Piezoresponse Force Microscopy and Kelvin Probe Force Microscopy), and demonstrate the emergence of ferrocatalysis, upon cooling at the paraelectric-ferroelectric phase transition of BaTiO3 single crystals.

References
[1] K. Cordero-Edwards, et al. J. Phys. Chem. C, 120, 24048 (2016).
[2] N. Domingo, et al. Phys. Chem.Chem.Phys, 21 (2019), 4920
[3] N.Domingo, et al., Nanoscale, 11 (2019) 17920
[4] I. Spasojevic, PhD Thesis, Universitat Autonoma de Barcelona, June 2022.

CONTROL ID: 4075648
CURRENT SYMPOSIUM: S3 - Advanced Electronic Materials: Processing and Structures and Properties and Applications
CURRENT SESSION: Advanced electronic materials, including ferroelectric, piezoelectric, dielectric, electrostrictive, electrocaloric and pyroelectric materials
TITLE: Scanning Oscillator Piezoresponse Force Microscopy:
new tools to explore domain wall dynamics.
PRESENTATION TYPE: Poster
PRESENTER (FIRST NAME ONLY): Neus
PRESENTER (LAST NAME ONLY): Domingo Marimon
PRESENTER (INSTITUTION ONLY): Oak Ridge National Lab
PRESENTER (EMAIL ONLY): ndomingo@ornl.gov
PRESENTER (COUNTRY ONLY): United States
DATE/TIME SUBMITTED: January 18, 2024, 09:39 PM
ABSTRACT BODY:
Abstract Body: The CNMS has developed Scanning Oscillator PFM as a novel tool to study dynamic fenomena in polar and electromechanically active materials. This microscopy mode is based on a multifrequency approach that combines slow and fast electric signal stimulus to resolve dynamic reversible phenomena in the second to sub-milisecond range, such as domain wall vibrational modes. Dynamics of ferroelectric domain walls associated to domain wall switching are known to depend on bulk structure, being very sensitive to defects, chemical and structural pinning sites, as well asenvironmental conditions, modifying the electronic boundary conditions associated with screening dynamics. However, domain walls also show a sub-coercive field dynamics as a reversible motion, with vibrational states that strongly couple to local structure and composition of the domain wall. This technique allows quick visualization of domain wall displacement, their velocities and dependence on pre-existing domain configurations and defects. When applied to lead titanate, this technique shows significant oscillations of the 180° domain walls between the antiparallel c+/c- domains.
This work was supported by Center for Nanophase Materials Sciences (CNMS), which is a US Department of Energy, Office of Science User Facility at Oak Ridge National Laboratory.

CONTROL ID: 4082899
CURRENT SYMPOSIUM: S3 - Advanced Electronic Materials: Processing and Structures and Properties and Applications
CURRENT SESSION: Advanced electronic materials, including ferroelectric, piezoelectric, dielectric, electrostrictive, electrocaloric and pyroelectric materials
TITLE: Controlling microstructure in relaxor ferroelectric $BaTiO_3$
PRESENTATION TYPE: Poster
PRESENTER (FIRST NAME ONLY): Vika
PRESENTER (LAST NAME ONLY): Sangwan
PRESENTER (INSTITUTION ONLY): Oregon State Univ
PRESENTER (EMAIL ONLY): sangwanv@oregonstate.edu
PRESENTER (COUNTRY ONLY): United States
DATE/TIME SUBMITTED: February 08, 2024, 03:22 PM
ABSTRACT BODY:
Abstract Body: The behavior of $BaTiO_3$ ceramics has been studied for ceramic capacitor applications. Lanthanide-doping and vacancy compensation have been used to create a shift towards a relaxor behavior in $BaTiO_3$ ceramics. Density, morphology & dielectric properties have been studied.

AUTHOR INDEX